P9-CJN-615

BIOTECHNOLOGY

A handbook of practical formulae

DIETER A. SUKATSCH

ALEXANDER DZIENGEL

Copublished in the United States with
John Wiley & Sons, Inc., New York

Longman Scientific & Technical,
Longman Group UK Limited,
Longman House, Burnt Mill, Harlow,
Essex CM20 2JE, England
and Associated Companies throughout the world.

Copublished in the United States with
John Wiley & Sons, Inc., 605 Third Avenue, New York, NY 10158

First published in German under the title *Formelsammlung Biotechnologie*
by BIBLIOMED-Medizinische Verlagsgesellschaft mbH, Melsungen 1984
This English translation first published 1987

British Library Cataloguing in Publication Data

Sukatsch, Dieter A.
Biotechnology: a handbook of practical
formulae.
1. Biotechnology 2. Science —— Formulae
I. Title II. Dziengel, Alexander
III. Formelsammlung Biotechnologie. *English*
660′.6′0212 TP248.2

ISBN 0-582-98899-3

Library of Congress Cataloging-in-Publication Data

Sukatsch, Dieter A.
Biotechnology: a handbook of practical formulae.

Translation of: Formelsammlung Biotechnologie.
Bibliography: p.
Includes index.
1. Biotechnology—Formulae—Handbooks, manuals,
etc. 2. Biotechnology—Nomenclature—Handbooks,
manuals, etc. I. Dziengel, Alexander. II. Title.
TP248.162.S8513 1987 660′.6 86-20826
ISBN 0-470-20729-9 (Wiley, USA only)

Set in 12/13pt Monophoto 2000 Times

Produced by Longman Group (FE) Limited
Printed in Hong Kong

Contents

Preface

This formulary is a summary of current nomenclature, definitions and associated equations currently in use in the field of biotechnology. It has not been attempted to extensively cover all aspects of microbiology, biochemical engineering, biophysics, molecular biology, physical chemistry and kinetics pertaining to the multidisciplinary field of biotechnology. It was, however, our aim to present a concise source of practical information. Therefore, terminology and those equations in more general use were given preference. Complex, specialized terminology has, as a consequence, not been included. We are aware that this is a compromise subject to criticism and any suggestions as well as any comments would be most welcome.

We would also like to thank Prof. R. M. Lafferty (Technical University, Graz, Austria) for having proof-read the English manuscript.

The authors

Prof. D. A. Sukatsch
HOECHST AG
P.O. Box 80
6230 Frankfurt
FRG

Dr A. Dziengel
B. BRAUN MELSUNGEN AG
P.O. Box 110
3508 Melsungen
FRG

MICROBIOLOGY

Growth kinetics – Number of cells
Discontinuous culture (batch)

$$N_t = N_0 \cdot 2^n$$

$$n = \frac{\log N_t - \log N_0}{\log 2}$$

$$v = \frac{n}{t} = \frac{1}{g} = \frac{\log N_t - \log N_0}{\log 2(t_t - t_0)}$$

N_t = number of cells at time t_t

N_0 = number of cells at time t_0

n = number of cell divisions

v = rate of cell division [h^{-1}]

g = generation time [h]

t = time [h]

Extent of validity: during the phase of exponential growth.

Growth kinetics – Biomass Discontinuous culture (batch)

$$\frac{dX}{dt} = \mu \cdot X \qquad X_t = X_0 \cdot e^{\mu t}$$

$$t_d = \frac{\ln 2}{\mu} \qquad \mu = \frac{\ln X_t - \ln X_0}{t_t - t_0}$$

X_t = biomass at time t_t [$g \cdot l^{-1}$]

X_0 = biomass at time t_0 [$g \cdot l^{-1}$]

t_d = doubling time of biomass [h]

μ = specific rate of growth [h^{-1}]

Extent of validity: during the phase of exponential growth, biomass is defined as dry material, often expressed as: [$g_{dm} \cdot l^{-1}$].

Growth kinetics – "Standard cells"

$$\mu = \nu \cdot \ln 2$$

$$t_d = g$$

μ = specific rate of growth [h^{-1}]

ν = rate of cell division [h^{-1}]

t_d = doubling time of biomass [h]

g = doubling time of cell number [h], equal to generation time

Extent of validity: for "standard cells", a doubling of the number of cells is equal to a doubling of biomass. This relationship holds true only during the phase of exponential growth.

Growth kinetics – Continuous culture Monod relationship

$$\mu = \mu_{max} \frac{S}{K_S + S}$$

μ = specific rate of growth $[h^{-1}]$

μ_{max} = maximum specific rate of growth $[h^{-1}]$

S = substrate concentration $[g \cdot l^{-1}]$

K_S = saturation constant $[g \cdot l^{-1}]$, substrate concentration at which $\mu = 0.5\,\mu_{max}$

The Monod relationship is only valid for special conditions analogous to those related to Michaelis–Menten-kinetics for enzyme activity (cf. enzyme kinetics). S is the concentration of the limiting substrate in the homogeneous continuous culture (chemostat).

Model equations for simple growth kinetics (no inhibition, no limitation)

$$\mu = \mu_{\max} \frac{S}{K_S + S}$$

$$\mu = \mu_{\max} [1 - e^{-S/K_s}]$$

$$\mu = \mu_{\max} \frac{S^n}{K_S + S^n}$$

$$\mu = \mu_{\max} \frac{S}{K_S \cdot X + S}$$

For explanation of symbols see previous pages.

Biomass determinations in technical media

$$\bar{X} = \frac{S_{ed} - \left(\frac{S_{ed}}{DW}\right)_0 \cdot DW}{\left(\frac{S_{ed}}{DW}\right)_1 - \left(\frac{S_{ed}}{DW}\right)_0}$$

$\bar{X}$ = average concentration of biomass

S_{ed} = sediment after filtration or centrifugation and drying $[g \cdot l^{-1}]$

DW = total dry mass of the media $[g \cdot l^{-1}]$

$0 \triangleq$ before inoculation

$1 \triangleq$ during course of bioprocess

This method is used for fermentation media containing insoluble substrates. The value of $(S_{ed} \cdot DW^{-1})_0$ must be determined only once before inoculation. Values for S_{ed} or DW can also be determined by quick methods such as infra-red or microwave drying.

Yield coefficients – Substrate

$$Y_{X/S}=\frac{\Delta X}{\Delta S}=\frac{X_t-X_0}{S_0-S_t}$$

$$Y_{S/X}=\frac{\Delta S}{\Delta X}=(Y_{X/S})^{-1}$$

ΔX = biomass increase $[g_{dm}\cdot l^{-1}]$

ΔS = substrate consumption $[g_s\cdot l^{-1}]$ or $[mol_s\cdot l^{-1}]$

$Y_{X/S}$ = yield coefficient with respect to substrate $[g_{dm}\cdot g_s{}^{-1}]$ or $[-]$, or $[g_{dm}\cdot mol_s{}^{-1}]$

$Y_{X/S}$ is often expressed as Y_S. To avoid incorrect interpretations both subscripts should be used. Reciprocal values can be expressed as either $(Y_{X/S})^{-1}$ or $1/Y_{X/S}$. When being used in a dimensionless form in other mathematical derivations it should be taken into consideration that different masses are involved, i.e. grams of dry mass and grams of substrate.

These rules also apply for all subsequent coefficients.

Yield coefficients – Oxygen

$$Y_{X/O_2} = \frac{\Delta X}{\Delta c_{O_2}} \qquad Y_{O_2/X} = (Y_{X/O_2})^{-1}$$

Δc_{O_2} = oxygen consumption
$[g_{O_2} \cdot l^{-1}]$ or $[mol_{O_2} \cdot l^{-1}]$

Y_{X/O_2} = yield coefficient for biomass with respect to oxygen $[g_{dm} \cdot mol_{O_2}{}^{-1}]$

The symbols $Y_{X/O}$ and Y_O are also used. It should be noticed that the symbol "O" refers to a gram atom of oxygen and that a different value must be taken into account.

Yield coefficients – Product formation

$$Y_{X/P} = \frac{\Delta X}{\Delta P} \qquad Y_{P/X} = (Y_{X/P})^{-1}$$

ΔP = product formation
$[g_P \cdot l^{-1}]$ or $[mol_P \cdot l^{-1}]$

$Y_{X/P}$ = yield coefficient for biomass with respect to product formation
$[g_{dm} \cdot mol_P{}^{-1}]$

See page 14 for further explanations.

Yield coefficients – Carbon dioxide

$$Y_{X/CO_2} = \frac{\Delta X}{\Delta c_{CO_2}} \qquad Y_{CO_2/X} = (Y_{X/CO_2})^{-1}$$

Δc_{CO_2} = carbon dioxide formation $[g_{CO_2} \cdot l^{-1}]$ or $[mol_{CO_2} \cdot l^{-1}]$

Y_{X/CO_2} = yield coefficient for biomass with respect to carbon dioxide produced $[g_{dm} \cdot mol_{CO_2}{}^{-1}]$

See page 14 for further explanations.

Yield coefficient – Energy

$$Y_{X/ATP} = \frac{\Delta X}{\Delta c_{ATP}} = Y_{X/S} \cdot \frac{M_S}{a}$$

$Y_{X/ATP}$ = yield coefficient for biomass with respect to energy [$g_{dm} \cdot mol_{ATP}^{-1}$]

Δc_{ATP} = amount of ATP formed [$mol \cdot l^{-1}$]

M_S = molar mass of energy yielding substrate [$g \cdot mol^{-1}$]

a = value which represents the number of moles of ATP produced from each mole of substrate

For the determination of $Y_{X/ATP}$ both the substrate and the metabolic pathway (calculation of "a") must be known. For aerobic growth conditions, the determination of "a" is very difficult.

Yield coefficients – Available electrons

$$Y_{X/ave^-} = \frac{Y_{X/S}}{ave^-}$$

Y_{X/ave^-} = yield coefficient for biomass with respect to available electrons $[g_{dm} \cdot mol_{e^-}{}^{-1}]$

ave^- = available electrons
= number of electrons per mole of substrate available for formation of biomass and conversion of energy per available mole of electrons $[mol_{e^-} \cdot mol_S{}^{-1}]$

$Y_{X/S}$ = yield coefficient with respect to substrate $[g_{dm} \cdot mol_S{}^{-1}]$

Yield coefficients – Heat

$$Y_{kcal^*} = Y_H = \frac{Y_{X/S}}{\Delta H_S - Y_{X/S} \cdot \Delta H_X}$$

Y_H = yield coefficient for biomass with respect to heat formation $[g_{dm} \cdot J^{-1}]$

$Y_{X/S}$ = yield coefficient for biomass with respect to substrate consumption $[g_{dm} \cdot g_S{}^{-1}]$

ΔH_S = heat of combustion of substrate $[J \cdot g_{dm}{}^{-1}]$

ΔH_X = heat of combustion of biomass formed $[J \cdot g_{dm}{}^{-1}]$

This relationship is based on the heat formed deriving only from microbial metabolism. It is necessary to know the substrate and the metabolic pathway. ΔH_X can be determined either by experiment or by calculation.

* The expression Y_{kcal} is no longer used.

Yield coefficients – Interrelationships

$$Y_{S/P} = \frac{\Delta S}{\Delta P} \qquad Y_{P/S} = \frac{\Delta P}{\Delta S}$$

$$Y_{CO_2/P} = \frac{\Delta c_{CO_2}}{\Delta P} \qquad Y_{P/CO_2} = \frac{\Delta P}{\Delta c_{CO_2}}$$

$$Y_{S/P} = \frac{\text{consumption of substrate in g or mol}}{\text{product formation in g or mol}}$$

$$Y_{CO_2/P} = \frac{\text{formation of } CO_2 \text{ in g or mol}}{\text{product formation in g or mol}}$$

Further yield coefficients can be formed in an analogous manner.

Metabolic rates – Oxygen uptake

$$Q_{O_2} = \frac{d(c_{O_2})}{dt}$$

$$q_{O_2} = \frac{1}{X} \cdot \frac{d(c_{O_2})}{dt} = \frac{Q_{O_2}}{X}$$

Q_{O_2} = volumetric oxygen uptake rate
$[g_{O_2} \cdot l^{-1} \cdot h^{-1}]$ or $[mol_{O_2} \cdot l^{-1} \cdot h^{-1}]$

q_{O_2} = specific oxygen uptake rate
$[g_{O_2} \cdot g_{dm}{}^{-1} \cdot h^{-1}]$ or $[mol_{O_2} \cdot g_{dm}{}^{-1} \cdot h^{-1}]$

The expressions of metabolic rates in capital and small letters have different meanings. The following equations and presentations will refer to the unit **mol** for substrates and products.

In addition to those metabolic rates to be further described, many others analogous to yield coefficients are in use such as:

Q_{ATP} or q_{ATP},

Q_{H} or q_{H},

Q_{ave^-} or q_{ave^-}, etc.

Metabolic rates – Oxygen uptake

$$Q_{O_2} = N_A = \frac{\mu \cdot X}{Y_{X/O_2}}$$

$Q_{O_2} = N_A =$ volumetric oxygen uptake rate $[\text{mol}_{O_2} \cdot l^{-1} \cdot h^{-1}]$

$\mu =$ specific growth rate $[h^{-1}]$

$Y_{X/O_2} =$ yield coefficient with respect to oxygen $[g_{dm} \cdot \text{mol}_{O_2}{}^{-1}]$

In continuous culture (chemostat)

$\mu \equiv D$

$D =$ dilution rate $[h^{-1}]$.

Metabolic rates – Carbon dioxide formation

$$Q_{CO_2} = \frac{d(c_{CO_2})}{dt}$$

$$q_{CO_2} = \frac{Q_{CO_2}}{X}$$

Q_{CO_2} = volumetric rate of carbon dioxide formation $[mol_{CO_2} \cdot l^{-1} \cdot h^{-1}]$

q_{CO_2} = specific rate of carbon dioxide formation $[mol_{CO_2} \cdot g_{dm}{}^{-1} \cdot h^{-1}]$

For explanations refer to page 22.

Metabolic rates – Biomass formation

$$Q_X = \frac{dX}{dt} \equiv \mu \cdot X$$

$$q_X = \frac{Q_X}{X} \equiv \mu$$

Q_X = volumetric rate of biomass formation $[g_{dm} \cdot l^{-1} \cdot h^{-1}]$

q_X = specific rate of biomass formation $[h^{-1}]$
= specific growth rate μ

For explanations refer to page 22.

Metabolic rates – Substrate uptake

$$Q_S = \frac{dS}{dt}$$

$$q_S = \frac{Q_S}{X}$$

Q_S = volumetric rate of substrate uptake $[mol_S \cdot l^{-1} \cdot h^{-1}]$

q_S = specific rate of substrate uptake $[mol_S \cdot g_{dm}{}^{-1} \cdot h^{-1}]$

For explanations refer to page 22.

Metabolic rates – Product formation

$$Q_P = \frac{dP}{dt}$$

$$q_P = \frac{Q_P}{X}$$

Q_P = volumetric rate of product formation $[mol_P \cdot l^{-1} \cdot h^{-1}]$

q_P = specific rate of product formation $[mol_P \cdot g_{dm}{}^{-1} \cdot h^{-1}]$

For explanations refer to page 22.

Respiratory quotient

$$RQ = \frac{Q_{CO_2}}{Q_{O_2}}$$

RQ = respiratory quotient

$$= \frac{\text{rate of carbon dioxide formation}}{\text{rate of oxygen uptake}}$$

Relationship between yield coefficients and metabolic rates

$$Y_{P/S} = \frac{Q_P}{Q_S}$$

$Y_{P/S}$ = product-substrate coefficient $[mol_P \cdot mol_S^{-1}]$

Q_P = rate of product formation $[mol_P \cdot l^{-1} \cdot h^{-1}]$

Q_S = rate of substrate uptake $[mol_S \cdot l^{-1} \cdot h^{-1}]$

Similar correlations can be obtained with other metabolic rates.

The following example is of special value:

$$Y_{X/O_2} = \frac{\mu}{q_{O_2}} \mathrel{\hat{=}} \frac{D}{q_{O_2}},$$ cf. page 30,

where D = dilution rate in continuous culture.

Relationships between yield coefficients

$$q_S = \frac{1}{X} \cdot \frac{dS}{dt} = \frac{dX}{dt} \cdot \frac{1}{X} \cdot \frac{dS}{dt} \cdot \frac{dt}{dX}$$

$$q_S = \mu \cdot Y_{S/X}$$

$$q_S = \frac{\mu}{Y_{X/S}}$$

$$Q_S = \frac{dS}{dt} = \frac{dS}{dt} \cdot \frac{dt}{dX} \cdot \frac{dX}{dt}$$

$$Q_S = Y_{S/X} \cdot \mu \cdot X$$

$$Q_S = \frac{\mu \cdot X}{Y_{X/S}}$$

For the quantitative determinations of other metabolic parameters, similar relationships can be formulated.

Substrate consumption – Batch growth

$$\frac{\mathrm{d}S}{\mathrm{d}t}=\frac{\mu X}{Y_{\mathrm{X/S}}}+mX+\frac{Q_{\mathrm{P}}}{Y_{\mathrm{P/S}}}$$

$\frac{\mathrm{d}S}{\mathrm{d}t}$ = decrease of substrate concentration with respect to time $[\mathrm{g_S} \cdot \mathrm{l}^{-1} \cdot \mathrm{h}^{-1}]$

$\frac{\mu X}{Y_{\mathrm{X/S}}}$ = fraction of substrate required for biomass formation $[\mathrm{g_S} \cdot \mathrm{l}^{-1} \cdot \mathrm{h}^{-1}]$

m = maintenance energy coefficient

$m \cdot X$ = fraction of substrate required for maintenance energy $[\mathrm{g_S} \cdot \mathrm{l}^{-1} \cdot \mathrm{h}^{-1}]$

$\frac{Q_{\mathrm{P}}}{Y_{\mathrm{P/S}}}$ = fraction of substrate required for product formation

Prerequisite: neither substrate nor product will be added to or removed from the system.

Growth with double substrate limitation

$$Q_X = \frac{dX}{dt} = \mu_{max} \cdot X \frac{S}{K_S + S} \cdot \frac{c_{O_2}}{K_{O_2} + c_{O_2}}$$

Q_X = rate of biomass formation $[g_{dm} \cdot l^{-1} \cdot h^{-1}]$

S = substrate concentration $[g_S \cdot l^{-1}]$

K_S = saturation constant $[g_S \cdot l^{-1}]$, substrate concentration at which $\mu = 0.5\, \mu_{max}$

c_{O_2} = oxygen concentration $[g_{O_2} \cdot l^{-1}]$

K_{O_2} = saturation constant $[g_{O_2} \cdot l^{-1}]$ at which $\mu = 0.5\, \mu_{max}$

S is often both the carbon and energy source.

Rate of microbial heat formation

$$Q_W = \frac{V_0 \cdot \mu \cdot X}{Y_H}$$

Q_W = rate of microbial heat formation $[J \cdot h^{-1}]$

V_0 = working volume [l]

Y_H = coefficient of microbial heat formation $[g_{dm} \cdot J^{-1}]$

Rate of product formation

$\frac{dP}{dt} = Q_P$	general
$\frac{dP}{dt} = \frac{\mu \cdot X}{Y_{X/P}}$	Gaden Type I
$\frac{dP}{dt} = \alpha \frac{dX}{dt} + \beta \cdot X$, or $q_P = \alpha \cdot \mu + \beta$	Gaden Type II and III

General equations according to Luedeking and Piret. In the Gaden Type II relationship, $\beta = 0$. In other cases, α and β are process-specific parameters which have to be experimentally determined.

Productivity of batch processes

$$P = \frac{X_t - X_0}{t_{tot}} = \frac{Y_{X/S} \cdot S_0}{\frac{1}{\mu_{max}} \cdot \ln \frac{X_t}{X_0} + t_1 + t_2 + t_3}$$

P = productivity of a batch culture with respect to biomass $[g_{dm} \cdot l^{-1} \cdot h^{-1}]$

X_t = concentration of biomass at harvesting time $[g_{dm} \cdot l^{-1}]$

X_0 = concentration of biomass after inoculation $[g_{dm} \cdot l^{-1}]$

$t_{tot} = t_{exp} + t_1 + t_2 + t_3$ = total time of process [h]

$$t_{exp} = \frac{1}{\mu_{max}} \cdot \ln \frac{X_t}{X_0} \quad [h]$$

t_1 = lag phase [h]

t_2 = time necessary for sterilization of fermentor and medium [h]

t_3 = time required for harvesting and cleaning of the fermentor and ancillary equipment

In many cases, $(t_2 + t_3) = t_l$, or $(t_1 + t_2 + t_3) = t_l$ is designated as "dead time".

Pasteur quotient

$$P_Q = \frac{Q_{S,a}}{Q_{S,an}}$$

P_Q = Pasteur quotient

$Q_{S,a}$ = rate of substrate uptake under aerobic conditions $[g_S \cdot l^{-1} \cdot h^{-1}]$

$Q_{S,an}$ = rate of substrate uptake under anaerobic conditions $[g_S \cdot l^{-1} \cdot h^{-1}]$

When $P_Q = 1$, this indicates that there is no regulatory effect present with respect to oxygen.

Growth kinetics – Continuous culture

$$\frac{dX}{dt} = \mu X - DX = X(\mu - D)$$

$$D = \frac{f}{V_0}$$

μ = specific growth rate [h^{-1}]

D = dilution rate [h^{-1}]

f = feed rate of medium = discharge rate of culture broth [l · h^{-1}]

V_0 = working volume [l]

Under steady-state conditions in the chemostat $\mu \equiv D$ and $dX/dt = 0$.

If $\mu > D$, the concentration of cells increases; if $\mu < D$, the concentration of cells decreases and eventually "washout" occurs.

Growth kinetics – Continuous culture

$$\frac{\mu \cdot X}{Y_{X/S}} = D(S_0 - S);$$

for $D = \mu$: $X = Y_{X/S}(S_0 - S)$;

for $S \ll S_0$: $X = Y_{X/S} \cdot S_0$

X = concentration of biomass during continuous culture in the chemostat $[g_{dm} \cdot l^{-1}]$

S_0 = concentration of substrate at inlet $[g_S \cdot l^{-1}]$

S = concentration of substrate at outlet $[g_S \cdot l^{-1}]$

$Y_{X/S}$ = yield coefficient with respect to substrate $[g_{dm} \cdot g_S^{-1}]$

D = dilution rate $[h^{-1}]$

The following prerequisites must be fulfilled:

X is independent of D,
X is only limited by S,
X is independent of all other components of the medium.

Growth kinetics – Continuous culture

$$S = \frac{D \cdot K_S}{\mu_{max} - D}$$

$$X = Y_{X/S}\left(S_0 - \frac{D \cdot K_S}{\mu_{max} - D}\right)$$

$$K_S = \frac{(\mu_{max} - D)S}{D}$$

K_S = saturation constant $[g_S \cdot l^{-1}]$, substrate concentration at which $\mu = 0.5\,\mu_{max}$

For explanation of symbols see previous page.

Product concentration – Continuous culture

$$P=\frac{Q_P}{D}=\frac{Q_P \cdot V_0}{f}$$

P = concentration of product in a homogeneously mixed chemostat $[g_P \cdot l^{-1}]$

Q_P = product formation rate $[g_P \cdot l^{-1} \cdot h^{-1}]$

D = dilution rate $[h^{-1}]$

V_0 = working volume [l]

f = feed rate of medium $[l \cdot h^{-1}]$

Contamination of a continuous culture

$$\frac{dX}{dt} = (\mu_X - D) \cdot X$$

$$\frac{dZ}{dt} = (\mu_Z - D) \cdot Z$$

X = concentration of microorganisms used in continuous culture

Z = concentration of the contaminating microorganisms

dX/dt = washout rate of biomass (X) in a continuous culture contaminated by Z

dZ/dt = increase rate of biomass of the contaminant

When $|dZ/dt| > 0$, μ_Z is larger than D and the culture becomes heavily contaminated. In the case of substrate limitation in a continuous culture, the value of dX/dt is always negative. Substrate concentration decreases because of the presence of contaminant Z and since D remains constant.

Single-stage, continuous fermentation with constant recycling of biomass and feeding of medium under sterile conditions ($X_0 = 0$)

$$\mu = D\left[1 + \alpha\left(1 - \frac{X_2}{X_1}\right)\right]$$

$$D = \frac{f_{fr}}{V_0} \qquad \alpha = \frac{f - f_{fr}}{f_{fr}}$$

μ = specific growth rate of total biomass [h^{-1}]

α = ratio of recycling

f_{fr} = feed rate of fresh medium [$l \cdot h^{-1}$]

f = flow rate of medium leaving the fermentor [$l \cdot h^{-1}$]

X_1 = concentration of cells in outflow [$g_{dm} \cdot l^{-1}$]

X_2 = concentration of cells in the recycle stream [$g_{dm} \cdot l^{-1}$]

$f_R = \alpha \cdot f_{fr}$ = recycle stream from separator to fermentor [$l \cdot h^{-1}$]

$$\left[1 + \alpha\left(1 - \frac{X_2}{X_1}\right)\right] < 1, \quad \text{therefore } \mu < D.$$

Continued on page 43.

Single-stage, continuous fermentation with constant recycling of biomass

$$X = \frac{Y_{X/S}(S_0 - S)}{1 + \alpha\left(1 - \frac{X_2}{X_1}\right)}$$

X = concentration of biomass during "steady-state" conditions $[g_{dm} \cdot l^{-1}]$

Since the numerator of the equation represents the increase of biomass in comparison with a culture without recycle, the following equation is valid:

$$A = \left\{\left[1 + \alpha\left(1 - \frac{X_2}{X_1}\right)\right]^{-1} - 1\right\} \cdot 100$$

A is the increase in yield (in %) of the process with recycle in comparison with the process without recycle.

Maximum yield in the chemostat

$$D_{opt} = \mu_{max}\left[1 - \sqrt{\frac{K_S}{K_S + S_0}}\right]$$

$$X_{max} = Y_{X/S}[S_0 - K_S - \sqrt{K_S(K_S + S_0)}]$$

D_{opt} = optimal dilution rate $\hat{=} \mu_{max}$ [h^{-1}]

X_{max} = maximum cell density at D_{opt} [$g \cdot l^{-1}$]

K_S = saturation constant [$g_S \cdot l^{-1}$], substrate concentration at which $\mu = 0.5\,\mu_{max}$

S_0 = concentration of substrate in inflowing medium [$g_S \cdot l^{-1}$]

Productivity of a continuous culture

$$P_{CC} = D \cdot X$$

$$P_{CC} = D \cdot Y_{X/S}\left(S_0 - \frac{D \cdot K_S}{\mu_{max} - D}\right)$$

P_{CC} = productivity of a continuous culture in the chemostat $[g_{dm} \cdot l^{-1} \cdot h^{-1}]$

D = dilution rate $[h^{-1}]$

$Y_{X/S}$ = yield coefficient with respect to substrate $[g_{dm} \cdot g_S{}^{-1}]$

Relationship between productivities of continuous and batch cultures

$$C = \frac{P_{CC,max}}{P_{batch}} = \frac{X_{max} \cdot D_{opt}}{P_{batch}}$$

$$C \approx \ln \frac{X_{max}}{X_0} + \mu_{max}(t_1 + t_2 + t_3)$$

C = relationship between maximum productivities of continuous and batch cultures

Symbols, see 35, 44 and 45

This relationship is approximately valid when $K_S \ll S_0$. At high growth rates, the value of "C" increases in favour of the continuous culture.

Specific microbial consumption of oxygen

$$|Y_{O_2/X}| = \frac{32 \cdot C + 8 \cdot H - 16 \cdot O}{|Y_{X/S}| \cdot |M_S|} + O' - 2.67 \cdot C' + 1.7 \cdot N' - 8 \cdot H'$$

$|Y_{O_2/X}|$ = numerical value of the specific oxygen consumption $[g_{O_2} \cdot g_{dm}{}^{-1}]$

C, H, O = number of atoms C, H, O, per molecule of substrate

$|Y_{X/S}|$ = numerical value of the yield coefficient with respect to substrate $[g_{dm} \cdot g_S{}^{-1}]$

$|M_S|$ = numerical value of the molar mass of substrate $[g_S \cdot mol^{-1}]$

O′, C′, N′, H′ = atomic fraction of the elements O, C, N, H in biomass, expressed as a fraction of 1.

This is an approximate equation for many aerobic processes; an elementary analysis of the biomass is necessary. Assumptions: cells, CO_2 and H_2O are the only reaction products; nitrogen source is ammonia.

BIOCHEMISTRY

Basic equations for enzyme kinetics according to Michaelis and Menten

$$E + S \underset{k_{-1}}{\overset{k_1}{\rightleftarrows}} ES \xrightarrow{k_2} E + P$$

$$v = \frac{v_{\max} S}{K_M + S}$$

$$v_{\max} = k_2 \cdot ES$$

$$K_S = \frac{k_{-1} + k_2}{k_1} \quad \text{and for } k_2 \ll k_{-1}$$

$$K_S = \frac{k_{-1}}{k_1} = K_M$$

E = enzyme or its concentration

S = substrate or its concentration

ES = enzyme–substrate complex or its concentration

v = velocity of enzymatic reaction

$v_{\max}$ = maximum velocity of enzymatic reaction

K_S = coefficient of dissociation of the ES complex

K_M = Michaelis–Menten constant

Units of enzyme activity

$$1\ \text{U} = \frac{1\ \mu\text{mol}_\text{S}}{\text{min}} = 16.67\ \text{nKat}$$

$$1\ \text{IU} = \frac{1\ \text{U}}{\text{mg}_\text{Pr}}$$

U = "unit" abbreviation of enzyme activity; 1 μmol substrate conversion per min

nKat = 10^{-9} Katal = 10^{-9} [mol · s^{-1}]

IU = "international unit", specific activity with respect to 1 mg protein (pure enzyme protein)

Kat is according to SI Standards; however, U and IU are still in use.

Enzyme kinetics: linearization of the Michaelis–Menten relationship

$$\frac{1}{v}=\frac{K_M}{v_{max}}\cdot\frac{1}{S}+\frac{1}{v_{max}} \quad (1)$$

$$v=-K_M\frac{v}{S}+v_{max} \quad (2)$$

v = velocity of reaction $[\mu mol \cdot mg^{-1} \cdot min^{-1}]$ or $[min^{-1}]$

S = concentration of substrate $[mol \cdot l^{-1}]$

K_M = Michaelis–Menten constant $[mol \cdot l^{-1}]$

v_{max} = maximum reaction velocity $[min^{-1}]$

(1) Lineweaver–Burk equation for plotting $1/v$ vs. $1/S$
(2) Eadie–Hofstee equation for plotting v vs. v/S

When the molar mass of the enzyme is known, then v or v_{max} can be expressed either as min^{-1} or s^{-1}.

Enzyme kinetics: competitive inhibition

$$v = \frac{v_{\max} \cdot S}{K_M\left(1 + \frac{I}{K_i}\right) + S}$$

$$m_1 = \frac{K_M\left(1 + \frac{I}{K_i}\right)}{v_{\max}}$$

$$m_2 = \frac{I}{v_{\max}}$$

$$m_3 = \frac{K_M}{K_i \cdot S \cdot v_{\max}}$$

v = reaction velocity [min^{-1}]

$v_{\max}$ = maximum reaction velocity [min^{-1}]

S = substrate concentration [$\mathrm{mol} \cdot \mathrm{l}^{-1}$]

K_M = Michaelis–Menten constant [$\mathrm{mol} \cdot \mathrm{l}^{-1}$]

I = inhibitor concentration [$\mathrm{mol} \cdot \mathrm{l}^{-1}$]

K_i = dissociation constant of the EI complex [$\mathrm{mol} \cdot \mathrm{l}^{-1}$]

m_1 = slope from plot of $1/v$ vs. $1/S$

m_2 = slope from plot of S/v vs. S

m_3 = slope from plot of $1/v$ vs. I

Enzyme kinetics: non-competitive inhibition

$$v = \frac{v_{max} \cdot S}{(K_M + S)\left(1 + \frac{I}{K_i}\right)}$$

$$m_1 = \frac{K_M\left(1 + \frac{I}{K_i}\right)}{v_{max}}$$

$$m_2 = \frac{\left(1 + \frac{I}{K_i}\right)}{v_{max}}$$

$$m_3 = \frac{S + K_M}{K_i \cdot S \cdot v_{max}}$$

Symbols see page 52.

Enzyme kinetics: only non-competitive inhibition present

$$v = \frac{v_{max} \cdot S}{K_M + S\left(1 + \frac{I}{K_i}\right)}$$

$$m_1 = \frac{K_M}{v_{max}}$$

$$m_2 = \frac{1 + \frac{I}{K_i}}{v_{max}}$$

$$m_3 = \frac{1}{K_i \cdot v_{max}}$$

Symbols see page 52.

Determination of optical activity: Biot's law of polarimetry

$$\alpha = [\alpha]_\lambda^T \cdot c \cdot d \cdot 10^{-2}$$

α = measured angle of optical rotation

$[\alpha]_\lambda^T$ = specific rotation of a pure, optically active substance at a given temperature and wavelength λ $[0.1 \cdot {}^\circ C \cdot m^2 \cdot g^{-1}]$

d = length of light path [m]

c = concentration of the dissolved substance $[g \cdot l^{-1}]$

The rotation of an optically active substance depends on the wavelength, temperature, pH, solvent, and several other factors. Under standard conditions the specific optical rotation of a substance is constant and can be used both to identify optically active substances and to determine the concentration of the compound in question.

Measurement of extinction or optical density: Lambert–Beer's law of photometry

$$\log \frac{I_0}{I} = E = \varepsilon \cdot c \cdot d$$

E = extinction

ε = molar extinction coefficient $[\mathrm{cm}^2 \cdot \mu\mathrm{mol}^{-1}]$

c = concentration of substance $[\mathrm{mol} \cdot \mathrm{l}^{-1}]$

d = length of light path [cm]

I_0 = intensity of the incident beam

I = intensity of the transmitted beam

The Lambert–Beer relationship is only valid for dilute solutions, and will depend on the physical and physiochemical parameters of the measuring system.

1. When using light absorption to determine the concentration of cells with a photometer, ΔE is not proportional to c. It is therefore necessary to dilute the sample to obtain the same ΔE value and then multiply this by the dilution factor.

2. In the fields of biochemistry and enzymology, many quantitative determinations are carried out using the "optical test" according to Warburg. The principle of this method is based on the absorbance of NADH (NADPH) at 340 nm which is formed as a reduced cofactor during enzyme–substrate reactions.

Characterization of gel chromatography separation processes

$$K_{av} = \frac{V_S - V_H}{V_t - V_H} \qquad K_d = \frac{V_S - V_H}{V_i}$$

$$V_i = V_N - V_H \qquad \text{HETP} = \frac{l_t}{N}$$

K_{av}, K_d = distribution coefficients

V_S = retention volume of the substance S [ml]

V_H = exclusion volume, high molecular [ml]

V_L = exclusion volume, low molecular [ml]

V_i = internal volume of gel particle [ml]

HETP = "Height Equivalent to a Theoretical Plate" [cm]

l_t = effective separation length of the column [cm]

N = number of theoretical separation stages

This relationship is valid for both upflow or downflow techniques.

Determination of molar mass and molecular radius with gel chromatography

$$K_d = -A \cdot \log M + B$$

$$K_d = -A' \cdot \log r + B'$$

K_d = coefficient of distribution of the molecular fraction

M = molar mass [$g \cdot mol^{-1}$]

r = Stokes radius of the molecule based on the assumption that it is of spherical form [nm]

A, B, A', B' = constants, derived from the graphical presentation

The substance to be determined is analysed together with a mixture of known standards. The determined values of K_d or K_{av} are to be plotted either against M or r.

Determination of the molar mass using measurements of osmotic pressure

$$M = \frac{c}{p_\pi} RT$$

M = molar mass $[\text{g} \cdot \text{mol}^{-1}]$

c = concentration $[\text{g} \cdot \text{l}^{-1}]$

p_π = osmotic pressure [bar]

R = gas constant

T = absolute temperature [K]

Determinations of molar mass are carried out with pure, diluted solutions of a particular substance. Measurements are done at different concentrations and extrapolation is carried out from $c \to 0$.

Determination of molar mass on the basis of sedimentation equilibrium

$$M = \frac{2RT \ln \frac{c_2}{c_1}}{\omega^2(1-\bar{v}\rho)(x_2{}^2 - x_1{}^2)}$$

M = molar mass [g·mol^{-1}]

c_1, c_2 = concentration of the substance at two different points in the centrifugation tube [g·l^{-1}]

x_1, x_2 = distance of these points from the centre of rotation [cm]

ω = angular velocity [s^{-1}]

$\bar{v}$ = partial specific volume of substance [cm^3·g^{-1}]

ρ = solvent density [g·cm^{-3}]

With this method only low angular velocities are necessary. Determination of the diffusion coefficient is not required.

Svedberg's law for the determination of molar mass in the gravitational field of an ultracentrifuge

$$s = \frac{dx}{x} \frac{1}{\omega^2 dt} \qquad M = \frac{sRT}{D(1 - \bar{v}\rho)}$$

s = sedimentation coefficient [s]

x = distance from centre of rotation [cm]

M = molar mass [$g \cdot mol^{-1}$]

D = diffusion coefficient [$cm^2 \cdot s^{-1}$]

$\bar{v}$ = partial specific volume of the molecule [$cm^3 \cdot g^{-1}$]

ρ = solvent density [$g \cdot cm^{-3}$]

ω = angular velocity [s^{-1}]

$S_{20,W}$ is the velocity of sedimentation at 20 °C in water, which for proteins varies from 1 to $200 \cdot 10^{-13}$ s. For proteins in watery solutions, $\bar{v}$ is approximately $0.74\ cm^3 \cdot g^{-1}$. For many proteins, $D_{20,W}$ varies from 1 to $20 \cdot 10^{-7}\ cm^2 \cdot s^{-1}$.

Manometric method according to Warburg

$$V_G = k' \cdot h$$

$$k' = \frac{1}{10}\left[V_g \frac{273}{T} + V_l \cdot \alpha\right]$$

V_G = quantity of gas consumed or evolved [$ml \cdot h^{-1}$]

k' = constant of reaction vessel [cm^2]

$h \triangleq$ pressure difference as measured on a manometer: total over a period of time [$cm \cdot h^{-1}$]

V_g = volume of the gas phase [ml]

V_l = volume of the liquid phase [ml]

α = Bunsen absorption coefficient [$ml_{gas} \cdot ml_l{}^{-1}$]

T = absolute temperature [K]

This is the manometric method for the determination of the gas consumption/evolution from biological material.

Energy charge

$$EC = \frac{c_{ATP} + 0.5\, c_{ADP}}{c_{ATP} + c_{ADP} + c_{AMP}}$$

EC = Energy Charge

ATP = Adenosin–5′–triphosphate

ADP = Adenosin–5′–diphosphate

AMP = Adenosin–5′–monophosphate

$0 \leqslant EC \leqslant 1.$

The term *EC* is a unit of the intracellular availability of energy, e.g. of energy-rich adenylates.

$EC = 1$ means only ATP is available,

$EC = 0$ means only AMP is available.

EC regulates single enzymes or metabolic pathways. Very high values of *EC* mean, for example, that the activity of catabolic enzymes is inhibited. With exponentially growing cells the *EC* value is about 0.8.

The exact determination of *EC* is very difficult since the concentrations of intracellular substances rapidly change during and after cell disruption.

GC content of DNA

$$GC = \frac{G + C}{G + C + A + T} \cdot 100$$

GC = relative content of G and C in DNA [%]

G = Guanine [mol]

C = Cytosine [mol]

A = Adenine [mol]

T = Thymine [mol]

The effect of temperature on DNA is determined by the GC ratio.

The "melting point" T_m of DNA is a function of GC, and can be photometrically determined using UV wavelengths. The GC ratio has a relevance for the taxonomic classification of microorganisms.

PHYSICAL CHEMISTRY

Concentration of solutions

$$n = \frac{m}{M} = c \cdot V = c_m \cdot m_{SV}$$

$$x_a = \frac{n_a}{\sum n_i} \qquad x_a' = x_a \cdot 100$$

n = number of moles [mol]

m = mass of solute [g]

M = molar mass [$g \cdot mol^{-1}$]

c = molarity [$mol \cdot l_S{}^{-1}$]; S = solution

V = volume of solution [l]

c_m = molality [$mol \cdot kg_{SV}{}^{-1}$]; SV = solvent

m_{SV} = mass of solvent [kg]

x_a = molar fraction of substance a

n_a = number of moles of substance a [mol]

$\sum n_i$ = number of moles of solution [mol]

x_a' = mole percentage of substance a in solution [%]

Absolute values of molarity are temperature-dependent, those of molality are not. In accordance with newer international usage, the term "mole" is not only applicable to molecules but also to certain groups, for instance, to carboxyl groups, ions, electrons, etc.

Law of mass action

$$A + B \underset{v_2}{\overset{v_1}{\rightleftarrows}} C + D$$

$$v_1 = k_1 \cdot c_A \cdot c_B$$

$$v_2 = k_2 \cdot c_C \cdot c_D$$

for $v_1 = v_2$: $$K_a = \frac{c_C \cdot c_D}{c_A \cdot c_B}$$

$c_{A,B,C,D}$ = concentration of substances A . . . D

v_1 = forward reaction rate

v_2 = reverse reaction rate

k_1, k_2 = specific reaction rate constants

K_a = equilibrium constant of the law of mass action

The law of mass action describes the equilibrium of a reversible chemical (biochemical) reaction: $v_1 = v_2$.

For first-order reactions, k has the unit time^{-1}; for second-order reactions time^{-1} · concentration^{-1}.

pH value

$$\mathrm{pH} = -\log \frac{a_{\mathrm{H_3O^+}}}{[a_{\mathrm{H_3O^+}}]_1};$$

simplified: $\mathrm{pH} = -\log [f \cdot c'_{\mathrm{H_3O^+}}]$

pH = pondus hydrogenii = "hydrogen ion exponent"

$a_{\mathrm{H_3O^+}}$ = activity of hydronium ions $[\mathrm{mol \cdot l^{-1}}]$

$[a_{\mathrm{H_3O^+}}]_1 = 1\ \mathrm{mol \cdot l^{-1}}$

f = coefficient of activity

$c'_{\mathrm{H_3O^+}}$ = antilogarithm of the hydronium ion concentration $[\mathrm{mol \cdot l^{-1}}]$

The pH value is the negative decadic logarithm of the absolute value of the hydronium ion activity expressed with the units $\mathrm{mol \cdot l^{-1}}$. $0 < f < 1$; $f \approx 1$ for diluted strong acids and bases; in the case of higher concentrations or poorly dissociated molecules f is considerably lower.

pK_a value

$pK_a = -\log K_a$

pK_a = negative decadic logarithm of the equilibrium constant of the law of mass action

The pK_a of a reaction corresponds to the pH value at which half of the molecules are undissociated; K_a is then equivalent to the dissociation constant of the electrolyte.

Redox potential

$$E = E_0 + \frac{R \cdot T}{z \cdot F} \ln \frac{a_{Ox}}{a_{Red}}$$

E = potential of redox reaction [V]

E_0 = standard potential of reaction [V]

z = number of electrons transferred

a_{Ox} = product of activities of oxidized products $[\mathrm{mol} \cdot \mathrm{l}^{-1}]$

a_{Red} = product of activities of reduced products $[\mathrm{mol} \cdot \mathrm{l}^{-1}]$

F = the Faraday ($96485\ \mathrm{A \cdot s \cdot mol^{-1}}$)

The Nernst equation expresses the voltage of redox reactions taking into consideration the actual concentration of the reactants. It is the basis of calculations pertaining to the equilibrium state of electrolytes.

Potential of standard pH electrodes

$E_H = -0.059\,16 \cdot pH$

E_H = Nernst voltage of the hydrogen electrode [V]

A change of one unit (1.0) of the pH value results in either an increase or decrease of 59.16 mV. This constant is valid under standard conditions of 25 °C and with an H_2 pressure of 1 bar.

Isoelectric point

$$IP = \frac{pK_1 + pK_2}{2}$$

IP = isoelectric point or pH value of an amphoteric electrolyte with two dissociable groups

If more than two dissociable groups are present, the IP is not necessarily the arithmetical mean value of all pK values.

Henderson–Hasselbalch equation

$$pH = pK_a + \log \frac{a_A}{a_{KA}}$$

a_A = activity of anions $[mol \cdot l^{-1}]$

a_{KA} = activity of non-dissociated molecules $[mol \cdot l^{-1}]$

This equation is used to calculate buffer capacities. In the case of $pH = pK$, then $a_A = a_{KA}$ (cf. pK_a value).

Determination of ionic concentrations

$$I = \frac{1}{2}\Sigma(c_i \cdot w_i^2)$$

I = ionic concentration, ionic strength [$mol \cdot l^{-1}$]

c_i = concentration of the ion [$mol \cdot l^{-1}$]

w_i = valency of the ion

The ionic concentration of a solution is correlated to the conductivity and has a significant influence on the processes of diffusion and distribution.

General gas equation for ideal gases

$$p \cdot V = n \cdot R \cdot T = \frac{m}{M} \cdot R \cdot T$$

p = pressure $[N \cdot m^{-2}]$

V = gas volume $[m^3]$

R = gas constant = $8.31\ J \cdot mol^{-1} \cdot K^{-1}$

n = number of moles of gas [mol]

m = mass of gas [g]

M = molar mass of gas $[g \cdot mol^{-1}]$

Gas laws for isobaric and isothermic changes of state

Isobaric change of state – Gay-Lussac's law

$$\frac{V_1}{T_1}=\frac{V_2}{T_2} \qquad V=V_0(1+\alpha \cdot t)$$

Isothermic change of state – Boyle–Mariotte's law

$$p_1 \cdot V_1 = p_2 \cdot V_2$$

V = volume of gas [m^3]

T = absolute temperature [K]

t = temperature [°C]

V_0 = standard volume at 0 °C

$\alpha = (273\ K)^{-1}$

p = pressure [$N \cdot m^{-2}$]

Reduction of measured gas volumes to S.T.P. (273.15 K, 1.013 bar)

$$V_0 = \frac{p}{p_0(1+\alpha t)} \cdot V = f \cdot V$$

V_0 = volume of gas at S.T.P. [l]

$p_0 = 1.013 \cdot 10^5 \ \mathrm{N \cdot m^{-2}} = 1.013$ bar = 760 torr = 1.0 atm

V = measured volume of gas [l]

α = coefficient of expansion; for ideal gases $\alpha = 0.003\,660\,8$ [$°C^{-1}$]

t = temperature [°C]

S.T.P. = Standard Temperature and Pressure.

Values of f are to be found in the corresponding tables of handbooks.

Solubility of gases in liquids

$$p_{A,g} = X_{A,l} \cdot H^{T}_{A,SV}$$

$p_{A,g}$ = partial pressure of gas A in the gaseous phase [bar]

$X_{A,l}$ = mole fraction of gas A in the liquid phase.

$H^{T}_{A,SV}$ = Henry's constant for gas A in the solvent SV at the temperature T [bar]

This equation represents Henry's law. According to the Henry–Dalton equation, it is also applicable for calculating the solubility of one component of a mixture of gases (air) in the liquid phase. For the solubility of oxygen in water, $H^{T}_{A,SV}$ is approximately $4 \cdot 10^4$ bar at room temperature.

Henry's constant is different from the Henry number (cf. page 79).

Solubility of gases in liquids

$$He = \frac{H^{T}_{A,SV}}{p} \cdot \frac{M_l}{M_g}$$

He = Henry's number

$H^{T}_{A,SV}$ = Henry's constant [bar]

p = pressure [bar]

M_l = average molar mass of the liquid phase [$g \cdot mol^{-1}$]

M_g = average molar mass of the gaseous phase [$g \cdot mol^{-1}$]

The dimensionless Henry's number has been proposed by Spalding. In textbooks, Henry's constant and Henry's number are sometimes incorrectly exchanged.

Solubility of oxygen in water

$$c_L = \frac{p_G}{H'}$$

c_L = oxygen concentration in air-saturated water $[mg \cdot l^{-1}]$

p_G = partial pressure of oxygen in the gaseous phase [bar]

H' = Henry's constant. At 25 °C: $H' = 0.026$ $[bar \cdot l \cdot mg_{O_2}{}^{-1}]$

H' values are more practical in use than $H^T_{A,SV}$. These are tabulated in different textbooks; α, Bunsen's absorption coefficients are also used. $p_G = 0.21$ bar only for dry air under standard conditions (cf. page 81). Instead of H', solubility coefficients can also be used.

Partial pressure of a gas (oxygen) in a humid gas mixture (air)

$$p_G = (p_B + p_O - p_L)\frac{A}{100}$$

p_G = partial pressure of gases [bar]

p_B = barometric pressure [bar]

p_O = (eventual) overpressure [bar]

p_L = partial pressure of steam [bar]

A = portion of gas in the dry gas mixture, in % vol. (for oxygen in air: $A = 21$)

When carrying out measurements of the oxygen partial pressure, at the lower depths in industrial fermentors, for instance, the effect of hydrostatic pressure must also be taken into consideration. However, in the absence of air bubbles, there is no effect of hydrostatic pressure on p_G.

Temperature effect on the solubility of oxygen in water

$$c_L = 14.16 - 0.394 \cdot t + 7.714 \cdot 10^{-3} t^2 - 6.46 \cdot 10^{-5} t^3$$

c_L = oxygen concentration in air-saturated water [$mg \cdot l^{-1}$]

t = temperature of water [°C]

Despite the decrease of oxygen solubility in water with increasing temperature, c_L is also dependent on the nature and concentration of other dissolved substances.

Increasing concentrations of dissolved carbohydrates reduce the value of c_L independent of the molar mass. In the case of proteins and/or peptides there is an additional effect of molar mass. The average value for the concentration of dissolved oxygen in technical fermentation media is 5–6 $mg_{O_2} \cdot l^{-1}$ at a temperature of 30 °C.

Thermodynamic equation for all chemical and biochemical reactions

$$\Delta G = \Delta H - T\,\Delta S$$

ΔG = change of free energy [$J \cdot mol^{-1}$]

ΔH = change of enthalpy [$J \cdot mol^{-1}$]

ΔS = change of entropy [$J \cdot mol^{-1} \cdot K^{-1}$]

This relationship is also applicable to enzyme kinetics and to the thermodynamic quantification of multi-step reactions.

Change of free energy

$A + B \rightleftarrows C + D$

$$\Delta G = \Delta G^\circ + RT \ln \frac{a_C \cdot a_D}{a_A \cdot a_B}$$

$$= \Delta G^\circ + RT \ln K_a$$

When $\Delta G = 0$ during the state of equilibrium, then

$\Delta G^\circ = -RT \ln K_a$

ΔG = change of free energy $[\mathrm{J \cdot mol^{-1}}]$

ΔG° = free energy change under standard conditions: pH 0.0, concentration of all substances 1 molar

K_a = equilibrium constant

In the case of biological reactions, $\Delta G^{\circ\prime}$ is used instead of ΔG°. $\Delta G^{\circ\prime}$ refers to a pH value of 7.0. K_a is then equal to K_a'. The influence of the concentration values of the reactants, that is intracellular compounds, on $\Delta G^{\circ\prime}$ is of great importance. If the value of ΔG° is negative, then the reaction occurs spontaneously once the activation energy level has been exceeded.

Arrhenius's law for quantification of sterilization processes

$$k = \text{const.} \cdot e^{-E/RT};$$

$$\ln k = -\frac{E}{RT} + \text{const.}$$

$$t = -\frac{1}{k} \cdot \ln \frac{N_t}{N_0}$$

k = reaction rate [s^{-1}] or equivalent correlating value

E = energy of activation; for thermoresistant spores $5 \cdot 10^5$ kJ $\cdot$ mol^{-1} (max).

T = sterilization temperature [K]

R = gas constant

t = time at temperature T [s]

N_t = number of spores in contaminated medium

N_0 = final number of spores (germs), usually 10^{-3} to 10^{-4}

E can be determined from a plot of $\ln k$ versus T^{-1}. The Arrhenius equation is of similar relevance for all chemical, physiochemical and biological reactions.

Arrhenius's (1) and van't Hoff's (2) laws for thermodynamic characterization of chemical and biochemical processes

(1) $d \ln k = \frac{E_A}{R} \frac{dt}{T^2}$

(2) $d \ln K = \frac{\Delta H}{R} \frac{dt}{T^2}$

$K = \frac{k_{+1}}{k_{-1}}$; $K_M = \frac{k_{-1}}{k_{+1}}$; $K = K_M{}^{-1}$

k = reaction rate constant

E_A = energy of activation

K = equilibrium constant for the formation of the enzyme–substrate complex

K_M = Michaelis–Menten constant

ΔH = change of enthalpy during formation of the enzyme–substrate complex

The prerequisite is that all enzyme molecules are bound as enzyme–substrate complexes; $k_2 \ll k_{-1}$.

Q_{10} value and energy of activation

$$E_A = \frac{R \cdot T_1 \cdot T_2}{T_2 - T_1} \ln \frac{k_2}{k_1}$$

$$E_A = \frac{R \cdot T_1 \cdot T_2}{10\,[\mathrm{K}]} \ln Q_{10}$$

$$Q_{10} = \frac{k_{t+10}}{k_t}$$

E_A = energy of activation [$J \cdot mol^{-1}$]

k_1, k_2 = rate constants of reaction at temperatures T_1 and T_2 [K]

Q_{10} = practical value for biochemical and biological reactions. Q_{10} is the quotient of the rate constants of a reaction taking place at T_1 and $T_1 + 10\,°C$.

Sterilization: decimal reduction time D_{10} and Z value

$$D_{10} = t_2 - t_1$$

A graphical plot of ln N versus time t applies to:

D_{10} = decimal reduction time [min]

t_1 = time for ln N_1

t_2 = time for ln N_2; $N_2 = 0.1 \cdot N_1$

N = number of surviving organisms

The D_{10} value is the time required to reduce the microbial population by 90 % (T = const.). The maximum value in the case of the average thermal resistance of bacillus spores is $D_{10, 121\,°C} \approx 5$ min.

The Z value represents the necessary increase of temperature in order to reduce the required time for death of the organisms by 90 %. Z values can be obtained from the point of intersection on the abscissa when the logarithm of the D value is plotted against temperature (thermal death curve).

Sterilization: *F* values and *L* values

$F = n \cdot D_{10} \qquad L = F^{-1}$

F = required period of time to destroy *all* spores (organisms) in a suspension [min], at 121 °C

$n = \log N$ = decimal factor

N = initial number of cells

L = lethality rate [min^{-1}]

n_r = real decimal factor

For reasons of safety, $n_r > n$ in actual practice. For example, a calculated risk of 1 % (1 contaminated batch in 100) would mean that $n_r = n + 2$.

Flow from closed vessels

$$\mathring{V} = K \cdot A \sqrt{2g \cdot h + 2\frac{p_1 - p_2}{\rho}}$$

$\mathring{V}$ = volumetric flow rate from the port of a closed vessel $[m^3 \cdot s^{-1}]$

A = port area $[m^2]$

g = acceleration due to gravity = $9.81\, m \cdot s^{-2}$

h = effective height of liquid measured between surface of liquid and geometric centre of port [m]

p_1 = pressure inside the vessel $[N \cdot m^{-2}]$

p_2 = pressure outside the vessel near the port $[N \cdot m^{-2}]$

K = constant [−]; for sharp edged outlets ≈ 0.6, for rounded outlets ≈ 0.98, for cylindrical pipes ≈ 0.82

ρ = density of liquid $[kg \cdot m^{-3}]$

Equation of continuity for incompressible liquids

$$\mathring{V} = A_1 \cdot v_1 = A_2 \cdot v_2 = \text{const.}$$

$\mathring{V}$ = constant volumetric flow rate in a piping system [$m^3 \cdot s^{-1}$]

A = cross-section of pipe [m^2]

v = flow rate [$m \cdot s^{-1}$]

Bernoulli's law

$$p+\frac{\rho \cdot v^2}{2}+\rho \cdot g \cdot h=\text{const.}$$

p = hydrostatic pressure $[\mathrm{N \cdot m^{-2}}]$

ρ = density $[\mathrm{kg \cdot m^{-3}}]$

v = flow rate $[\mathrm{m \cdot s^{-1}}]$

h = height between the centre of flow and a reference level [m]

The total energy attributable to the flowing of a liquid is constant and consists of kinetic energy, energy due to pressure difference and potential energy.

Hagen–Poiseuille's law

$$\Delta p = \frac{128 \cdot \mathring{V} \cdot \eta \cdot l}{\pi \cdot d^4}$$

Δp = pressure drop, laminar flow of incompressible liquid in a pipe [$N \cdot m^{-2}$]

$\mathring{V}$ = flow rate [$m^3 \cdot s^{-1}$]

η = dynamic viscosity [$N \cdot s \cdot m^{-2}$]

l = length of pipe [m]

d = internal diameter of pipe [m]

Viscosity data

$$\eta_N = \frac{\tau}{D} = \tau \cdot \frac{dy}{dv}$$

$$\nu = \frac{\eta_N}{\rho}$$

$$\eta_R = \frac{\eta}{\eta_0} \qquad \eta_R{}' = \frac{\eta - \eta_0}{\eta_0}$$

η_N = dynamic viscosity of a liquid with Newtonian properties $[N \cdot s \cdot m^{-2}] = [kg \cdot m^{-1} \cdot s^{-1}] =$ [10 poise]

η = the force required to move the parallel surfaces of a liquid element of 1 cm^2 cross-sectional area and of 1 cm thickness with a relative speed of 1 $cm \cdot s^{-1}$

τ = shear stress $[N \cdot m^{-2}]$

D = shear rate $[s^{-1}]$

dy/dv = velocity gradient of shear

ν = kinematic viscosity $[m^2 \cdot s^{-1}] = [10^4$ stokes]

ρ = density of liquid $[kg \cdot m^{-3}]$

η_R = relative viscosity of a solution

$\eta_R{}'$ = relative change of viscosity of a solution

η_0 = solvent viscosity

Viscosity data

$$\tau = \eta \cdot \left(\frac{dv}{dy}\right)^n$$

τ = shear stress

η = dynamic viscosity = consistency index

n = flow behaviour index

dv/dy = shear rate

$n = 1$ Newtonian liquid

$n < 1$ structural viscosity

Fermentation broths for the growth of bacteria and yeast usually behave like Newtonian liquids at the beginning of the fermentation process. The growth of filamentous fungi or streptomycetes or the formation of extracellular polysaccharides results in changes of the rheological properties of the fermentation broth which no longer behaves as a Newtonian liquid.

Determination of viscosity with the Ostwald or Ubbelohde viscosimeter

$$\eta = C \cdot \rho \cdot \Delta t$$

$$C = \frac{\pi \cdot r^4 \cdot g \cdot \Delta h}{8 \cdot \Delta V \cdot l}$$

C = constant dependent on the geometry of the viscosimeter $[m^2 \cdot s^{-2}]$

r = internal radius of the capillary tube [m]

g = acceleration due to gravity = $9.81\,m \cdot s^{-2}$

Δh = average difference of liquid levels [m]

ΔV = liquid volume $[m^3]$

l = length of capillary tube [m]

C can also be experimentally determined by employing a liquid with a known viscosity. This method for the determination of liquid viscosity is only applicable for systems not containing particles.

Determination of viscosity with the Ostwald or Ubbelohde viscosimeter

$$\eta_R = \frac{\rho \cdot \Delta t}{\rho_0 \cdot \Delta t_0}$$

η_R = relative viscosity of a solution

ρ = solution density $[kg \cdot m^{-3}]$

ρ_0 = solvent density $[kg \cdot m^{-3}]$

Δt = time required for solution to flow through a defined length of capillary tube [s]

Δt_0 = time required for solvent to flow through a defined length of capillary tube [s]

No insoluble substances may be present in the liquid phase.

Determination of liquid viscosities using a falling-sphere type of viscometer

$$\eta = K(\rho_1 - \rho_2)\,\Delta t$$

K = constant of the solid sphere

ρ_1 = density of the sphere [$kg \cdot m^{-3}$]

ρ_2 = density of the solution [$kg \cdot m^{-3}$]

Δt = time required for the sphere to pass through a defined length of a precision tube [s]

The values of K and ρ_1 are listed in the manufacturer's instruction manual.

Determination of liquid viscosities using a rotating viscosimeter

$$\eta = K \cdot U \cdot S$$

η = dynamic viscosity $[\mathrm{N \cdot s \cdot m^{-2}}]$

K = constant of the rotating cylinder used $[\mathrm{N \cdot s \cdot m^{-2} \cdot Sc^{-1}}]$

U = predetermined range of velocity corresponding to a definite speed of rotation

S = measured scale value on viscosimeter [Sc]

This derivation applies to only one of a large number of commonly used measuring devices employing rotating bodies. Many devices based on the principle of measuring viscosity with rotating cylinders or other geometric forms such as cones are commercially available.

Determination of radiant energy

$$E = h \cdot v = h \frac{c}{\lambda}$$

E = energy of electromagnetic radiation [J]

h = Planck's constant [J · s⁻¹]

v = frequency of radiation [s⁻¹]

c = velocity of light [m · s⁻¹]

λ = wavelength [m]

This equation is useful in thermodynamics and for calculating the yield of phototrophic organisms.

Water activity and growth of microorganisms

$$a_W = \frac{p}{p_0}$$

a_W = water activity, a_W value

p = vapour pressure of water in the substrate

p_0 = vapour pressure of pure water

For most microorganisms growth occurs between $0.5 < a_W < 1.0$.

Minimum values for a_W are:

normal bacteria	0.91
normal yeasts	0.81
normal fungi	0.80
osmophilic yeast	0.60

The value of a_W is of importance in connection with the use of drying processes for the preservation of food at low a_W values.

BIOCHEMICAL ENGINEERING

Oxygen uptake rate

$$Q_{O_2} = \mathrm{OUR} = \frac{F_I}{V_0}(c_I - c_O) = \frac{\mu \cdot X}{Y_{X/O_2}}$$

Q_{O_2} = Oxygen Uptake Rate [$\mathrm{mol \cdot l^{-1} \cdot h^{-1}}$]

F_I = flow rate of inlet gas [$\mathrm{l \cdot h^{-1}}$]

V_0 = working volume [l]

c_I = O_2 concentration at gas inlet [$\mathrm{mol \cdot l^{-1}}$]

c_O = O_2 concentration at gas outlet [$\mathrm{mol \cdot l^{-1}}$]

Y_{X/O_2} = yield coefficient with respect to oxygen [$\mathrm{g_{dm} \cdot mol_{O_2}{}^{-1}}$]

The validity of this equation is based on the assumption that constant values exist for temperature, pressure, and flow rates of gases entering and leaving the fermentor. This method is a simplified determination of oxygen uptake rates under actual working conditions.

Oxygen uptake rate

$$Q_{O_2} = \frac{F_I}{V_M \cdot V_0} \left[X^I_{O_2} - \frac{1 - X^I_{O_2} - X^I_{CO_2}}{1 - X^O_{O_2} - X^O_{CO_2}} X^O_{O_2} \right]$$

Q_{O_2} = oxygen uptake rate [mol · l⁻¹ · h⁻¹]

F_I = flow rate of inlet gas [l · h⁻¹]

V_M = molar volume of gases = 22.4 [l · mol⁻¹]

V_0 = working volume, liquid phase [l]

$X^I_{O_2}$ = molar fraction of oxygen at gas inlet

$X^O_{O_2}$ = molar fraction of oxygen at gas outlet

$X^I_{CO_2}$ = molar fraction of CO_2 at gas inlet

$X^O_{CO_2}$ = molar fraction of CO_2 at gas outlet

This equation is based upon the inert gas balance. It is assumed that CO_2 is the only gaseous product of the fermentation.

Carbon dioxide formation rate

$$Q_{CO_2} = \frac{F_I}{V_M \cdot V_0} \left[X^O_{CO_2} \frac{1 - X^I_{O_2} - X^I_{CO_2}}{1 - X^O_{O_2} - X^O_{CO_2}} - X^I_{CO_2} \right]$$

Q_{CO_2} = carbon dioxide formation rate [mol·l^{-1}·h^{-1}]

Refer to symbols and explanations on page 104.

Oxygen uptake rate

$$Q_{O_2} = \frac{1}{V_0}\left[\frac{F_I \cdot p_I \cdot X_I}{R \cdot T_I} - \frac{F_O \cdot p_O \cdot X_O}{R \cdot T_O}\right]$$

I = gas inlet

O = gas outlet

Q_{O_2} = oxygen uptake rate [$\text{mol} \cdot \text{l}^{-1} \cdot \text{h}^{-1}$]

V_0 = working volume, liquid phase [l]

F = flow rate of gases [$\text{l} \cdot \text{h}^{-1}$]

p = total gas pressure [$\text{N} \cdot \text{m}^{-2}$]

X = oxygen content of gas in % of volume (100 % $\triangleq$ 1.0)

R = gas constant, 8.31 [$\text{N} \cdot \text{m} \cdot \text{mol}^{-1} \cdot \text{K}^{-1}$]

T = absolute temperature [K]

The relationship presented here takes into consideration the fact that F_I and F_O, as well as the physical parameters at the inlet and exhaust of the fermentor, may be different.

Oxygen transfer rate – Laboratory scale

$$\mathrm{OTR} = k_L \cdot a(c^* - c_L)$$

OTR = Oxygen Transfer Rate [$\mathrm{mol \cdot l^{-1} \cdot h^{-1}}$]

k_L = mass transfer coefficient for oxygen in the gas–liquid film interface [$\mathrm{m \cdot h^{-1}}$]

a = specific gas–liquid interfacial area per unit volume [$\mathrm{m^{-1}}$]

c^* = liquid phase concentration of oxygen in equilibrium with the gaseous phase (saturation conditions) [$\mathrm{mol \cdot l^{-1}}$]

c_L = liquid phase concentration of oxygen [$\mathrm{mol \cdot l^{-1}}$]

Synonyms for OTR are n_a and N_a.

This relationship is only valid when V_0 is homogeneously mixed and c^* is constant with respect to the complete contents of the fermentor. This equation is also applicable for $k_L \cdot a$ determinations using the "direct method".

Oxygen transfer rate – Production scale

$$\mathrm{OTR} = k_\mathrm{L} \cdot a \cdot \frac{(c_\mathrm{I}^* - c_\mathrm{L}) - (c_\mathrm{O}^* - c_\mathrm{L})}{\ln \dfrac{c_\mathrm{I}^* - c_\mathrm{L}}{c_\mathrm{O}^* - c_\mathrm{L}}}$$

c_I^* = liquid phase concentration of oxygen in equilibrium with inflowing gas stream [$\mathrm{mol} \cdot \mathrm{l}^{-1}$]

c_O^* = liquid-phase concentration of oxygen in equilibrium with exit gas [$\mathrm{mol} \cdot \mathrm{l}^{-1}$]

In the case of this equation, the average logarithmic concentration profile of oxygen in production scale fermentors is taken into consideration.

For explanation of other symbols see previous pages.

Specific gas interfacial area and relative gas hold-up

$$a = \frac{A}{V_0} = \frac{6\varepsilon}{d_B(1-\varepsilon)}$$

$$\varepsilon = \frac{V_G}{V_L + V_G} \qquad d_B = \frac{\sum_{i=1}^{n} (N_i d_i^3)}{\sum_{i=1}^{n} (N_i d_i^2)}$$

a = specific gas–liquid interfacial area per unit volume $[m^{-1}]$

A = total surface area of all gas bubbles $[m^2]$

V_0 = working volume, liquid phase $[m^3]$

ε = gas hold-up

d_B = average "Sauter" bubble diameter [m]

V_G = volume of gas in the mixture $[m^3]$

V_L = volume of liquid in the mixture $[m^3]$

N_i = number of bubbles with the diameter d_i

In actual fermentation media which are aerated, gas bubbles may not be present in the form of perfect spheres but as ellipsoids. Thus the actual value of a may be somewhat larger than the calculated value.

Determination of the $k_L \cdot a$ value

$$\frac{dc_L}{dt} = k_L \cdot a(c^* - c_L) - Q_{O_2}$$

dc_L/dt = change of the concentration of oxygen in the liquid phase with respect to time

This relationship is used for the determination of $k_L \cdot a$ values when using the "dynamic method". c_L is the only value which must be measured.

Explanation of symbols is given on page 107.

Aeration number

$$N_a = \frac{F_G}{N \cdot D_i^3} = \frac{v_S}{v_{tip}}$$

N_a = aeration number

F_G = volumetric gas flow rate $[m^3 \cdot s^{-1}]$

N = impeller speed $[s^{-1}]$

D_i = impeller diameter [m]

v_S = superficial gas velocity $[m \cdot s^{-1}]$ (see p. 113)

v_{tip} = impeller tip speed $[m \cdot s^{-1}]$ (see p. 120)

N_a together with the ratio P_G/P permits the evaluation of power input and the effect of the impeller geometry during aeration of fermentation liquids.

Oxygen demand during cellular growth

$$C = \frac{A}{Y_{X/S}} - B$$

C = oxygen requirement to produce one gram of cell dry mass $[g_{O_2} \cdot g_{dm}^{-1}]$

A = oxygen requirement for the complete combustion of one gram of substrate $[g_{O_2} \cdot g_S^{-1}]$

B = oxygen requirement for the complete combustion of one gram of cell dry mass $[g_{O_2} \cdot g_{dm}^{-1}]$

$Y_{X/S}$ = yield coefficient with respect to substrate $[g_{dm} \cdot g_S^{-1}]$

This is an approximation which is only relevant when constant ecological conditions are present for the population of cells.

Superficial gas velocity in fermentors

$$v_S = \frac{F_G}{A}$$

v_S = superficial gas velocity $[m \cdot s^{-1}]$
F_G = volumetric gas flow rate $[m^3 \cdot s^{-1}]$
A = free cross-sectional area of the fermentor $[m^2]$

In aerated fermentations, F_G is often related to the working volume:

$$F' = \frac{F_G}{V_0}$$

The usual designation for F' is VVM = Volume of gas per Volume of liquid per Minute.

Efficiency factor of aeration systems

$$\eta_{O_2} = \frac{c_{IN} - c_{OUT}}{c_{IN}} \cdot 100$$

η_{O_2} = oxygen efficiency factor [%]

c_{IN} = oxygen concentration of inlet gas

c_{OUT} = oxygen concentration of outlet gas

This is a basic calculation employed to evaluate both the efficiency and economy of operation of aeration systems.

Mixing efficiency in fermentors

$$m = \frac{c_\infty - c}{c_\infty - c_0} \cdot 100$$

m = mixing efficiency in a fermentor [%]

c = concentration of test substance in the medium

c_∞ = concentration of test substance after complete mixing

c_0 = initial concentration of test substance

Mixing efficiency is of basic importance for the determination of mixing times and is expressed as 5 % or 1 % deviation from the infinite value of perfect homogeneity. Thus it is equivalent to the condition when either 95 % or 99 % of the state of optimal mixing has been attained.

Mixing time (t_m or θ) can therefore be defined as the period of elapsed time between the initial addition of the test substance and the time when the value of m has been reached.

Power input during non-aerated fermentations

$$P = N_P \cdot \rho \cdot N^3 \cdot D_i^5$$

$$P^* = \frac{P}{V_0}$$

P = power input (consumption) in non-aerated fermentations [W]

P^* = specific power input [$W \cdot m^{-3}$]

N_P = power number (constant) of the impeller

ρ = liquid density [$kg \cdot m^{-3}$]

N = impeller speed [s^{-1}]

D_i = impeller diameter [m]

The power number (N_P) is related to the Reynolds and Froude numbers. It is also dependent on vessel and impeller geometry and can vary by a factor of up to 100 (cf. page 119).

Power input during aerated fermentations

$$P_G = K' \left[\frac{P^2 \cdot N \cdot D_i^3}{F_G^{0.56}} \right]^{0.45}$$

$K' = 1.935 \cdot C$; C is a geometric constant for the fermentor being investigated and 1.935 is the conversion factor from metric horsepower to watts ($1.935 = 735.5^{0.1}$)

P_G = power input under aerated conditions; numerical value in watts

P = power input under non-aerated conditions; numerical value in watts

N = impeller speed; numerical value in $\min^{-1}$

D_i = impeller diameter; numerical value in dm

F_G = volumetric gas flow rate; numerical value in $1 \cdot \min^{-1}$

This equation is an empirical relationship, which takes into consideration the contribution or effect of the relevant parameters influencing power input. This relationship is applicable to all flow characteristics with Newtonian liquids as well as to the turbulent region of non-Newtonian liquids. Good correlations are found for working volumes of up to $30\,m^3$.

Power factor in relation to oxygen transfer rate

$$L_{O_2} = \frac{P}{V_0 \cdot \mathrm{OTR}}$$

L_{O_2} = power factor, with respect to oxygen consumption $[\mathrm{kW \cdot h \cdot kg_{O_2}}^{-1}]$

P = power input to fermentors [kW]

V_0 = working volume, liquid phase $[\mathrm{m^3}]$

OTR = oxygen transfer rate of bioreactor $[\mathrm{kg_{O_2} \cdot m^{-3} \cdot h^{-1}}]$

Power number for fermentors

$N_P = K \cdot Re^{-a'}$

N_P = power number

K = constant, dependent on the geometry of the fermentor and the impeller

Re = Reynold's number (cf. page 140)

a' = exponent, dependent on flow characteristics

$0 \leq Re < 10$ (laminar): $a' = 1$

$10 < Re < 10^3$ (intermediate)

$10^3 < Re$ (turbulent): $a' = 0$

This correlation is suitable for scale-up calculations, since K is independent of the size of the vessel. N_P can be calculated from power input (cf. page 116).

Impeller tip speed

$$v_{tip} = \pi \cdot D_i \cdot N$$

v_{tip} = impeller tip speed [$m \cdot s^{-1}$]

D_i = maximum impeller diameter [m]

N = impeller rotational speed [s^{-1}]

Momentum factor and impeller factor

$$M_f = N^2 \cdot D_i \cdot W \cdot L(D_i - W) = N^2 \cdot D_i \cdot J_f$$

$$J_f = W \cdot L(D_i - W)$$

M_f = momentum factor [$m^4 \cdot s^{-2}$]

J_f = impeller factor [m^3]

N = impeller speed [s^{-1}]

D_i = maximum diameter of the impeller [m]

L = impeller blade length [m]

W = impeller blade width [m]

The correlation is only relevant for disc turbine impellers. M_f and J_f could be useful to calculate $k_L \cdot a$ values; their use has also been proposed for scale-up procedures. Standard impeller proportions are $D_i:L:W = 20:5:4$.

Recommendations for optimal spacing of multiple impellers

$$D_i < s < 2\,D_i \qquad s_2 = \frac{H_L - 0.9\,D_i}{2}$$

$$\frac{H_L - D_i}{D_i} > m > \frac{H_L - 2\,D_i}{D_i}$$

D_i = impeller diameter [m]

s = distance between the single impellers [m]

s_2 = distance between two impellers [m]

H_L = height of liquid in the fermentor [m]

m = number of impellers

Optimal conditions with respect to the number and spacing of multiple impellers must be experimentally determined while taking into consideration all microbiological and physical parameters of the specific fermentation (flow characteristics, effects of shear rate, etc.).

Spacing of *m* impellers on one shaft

$$s_m = \frac{\log m}{2A}\left[H_L - \left(0.9 + \frac{A}{\log m}\right)D_i\right]$$

$$A = \log m(m-1) - \log (m-1)!$$

s_m = distance between two impellers [m]

After s_m has been determined, the impellers will be symmetrically fitted to the shaft.

For explanation of other symbols see previous page.

Baffle ratio

$$BR = \frac{n_B \cdot W_B}{D}$$

BR = baffle ratio

n_B = number of baffles

W_B = width of baffles

D = internal diameter of the fermentor vessel

Heat transfer in heat exchangers

$$\mathring{Q} = k \cdot A \frac{\Delta t_1 - \Delta t_2}{\ln \frac{\Delta t_1}{\Delta t_2}} = k \cdot A \cdot \Delta t_m$$

$\mathring{Q}$ = quantity of heat transferred [W]
k = heat transfer coefficient $[W \cdot m^{-2} \cdot K^{-1}]$
A = heat transfer area (effective heating or cooling surface) $[m^2]$
Δt_1 = temperature difference with respect to medium at the beginning of the heat exchanger area [K]
Δt_2 = temperature difference with respect to medium at the end of the heat exchanger area [K]
Δt_m = average logarithmic temperature difference

This formula applies to both parallel and countercurrent flow; Δt_1 is often called the “big” and Δt_2 the “small” temperature difference.

Components of the heat transfer coefficient

$$k=\left[\frac{1}{\alpha_1}+\sum_{i=1}^{n}\frac{\delta_i}{\lambda_i}+\frac{1}{\alpha_2}\right]^{-1}$$

k = heat transfer coefficient $[W \cdot m^{-2} \cdot K^{-1}]$
Value for insulated piping in closed systems and with $\Delta t = 80$ K: $k = 5$
Value for heating tube coil, steam 8 bar/turbulent water: $k = 5\,000$

$\alpha_{1,2}$ = heat transmission coefficients $[W \cdot m^{-2} \cdot K^{-1}]$ for both sides of the heat transfer surfaces; typical values: static air: 3–30; agitated liquids: 200–3 000; condensing steam: 8 000–20 000

λ_i = thermal conduction coefficients $[W \cdot m^{-1} \cdot K^{-1}]$ of multilayered walls, also wall + coating; for steel (glass) at 100 °C: 15–50 (0.75)

δ_i = thickness of different layers [m]

Since k is always smaller than the smallest value of α, *both* values of α have to be optimized to attain maximum heat transfer.

Determination of the amount of condensate

$$\mathring{M} \approx 2.2 \cdot 10^{-3} |\mathring{Q}|$$

$\mathring{M}$ = flow of condensate [$kg \cdot h^{-1}$]

$\mathring{Q}$ = quantity of heat transferred, numerical value in watts

This approximation includes a factor for radiation of about 20 % and a heat of vaporization of 2 000 $kJ \cdot kg^{-1}$; the heat transfer medium is saturated steam. Principally the heat of vaporization is taken into consideration.

Determination of heat transfer from condensate flow

$$\mathring{Q} = \mathring{M} \cdot c_P \cdot \Delta t_m$$

$\mathring{Q}$ = quantity of heat transferred $[kJ \cdot h^{-1}]$

$\mathring{M}$ = condensate flow $[kg \cdot h^{-1}]$

c_P = specific heat of the medium $[kJ \cdot kg^{-1} \cdot K^{-1}]$

Δt_m = average logarithmic temperature difference [K] (cf. page 125)

Heat formation during a fermentation

$$Q_H = Q_F + Q_M + Q_V + Q_G + Q_R$$

Q_H = total heat formation [W]

Q_F = heat of fermentation

Q_M = heat from mixing

Q_V = heat from evaporation

Q_G = heat from gas stream

Q_R = heat radiation from the fermentor itself

In technical fermentations, Q_V and Q_G are usually negligible.

Specific heat formation during fermentations

$$Q_F' = \frac{\Delta T}{\Delta t} \cdot c_P$$

Q_F' = specific heat formation of fermentation [$W \cdot kg^{-1}$]

ΔT = temperature increase [K]

Δt = period of time [s]

c_P = specific heat capacity of the fermentor [$J \cdot kg^{-1} \cdot K^{-1}$]

For the determination of heat formation, the temperature regulator is switched off for a short time and the temperature increase (ΔT) is measured.

Heat formation during aerobic fermentations

$$|\mathring{Q}| \approx 3.3 \cdot |Q_{O_2}|$$

$\mathring{Q}$ = heat formation during aerobic fermentations; numerical value in $W \cdot l^{-1}$

Q_{O_2} = volumetric oxygen uptake rate, numerical value in $mg_{O_2} \cdot l^{-1} \cdot h^{-1}$

Oxygen transfer and scale-up strategies

$$k_L \cdot a = K\left[\left(\frac{P_G}{V_0}\right)^a \cdot v_S{}^b\right]$$

$$k_L \cdot a = K(N^3 \cdot D_i{}^2)$$

P_G/V_0 = specific power input of gassed fermentations $[W \cdot m^{-3}]$

v_S = superficial gas velocity $[m \cdot s^{-1}]$; ratio between volumetric gas flow and free cross-sectional area of the fermentor

a,b = exponents, practically constant for one type of impeller and geometrically similar fermentor vessels

N = impeller speed $[s^{-1}]$

D_i = impeller diameter $[m]$

K = proportionality constant

These are equations for scale-up strategies of gassed fermentations. The average bubble velocity, liquid density, surface aeration, and energy losses are regarded as being constant.

Oxygen transfer and scale-up strategies

$$k_L \cdot a = K\left(\frac{P_G}{V_0}\right)^{0.4} \cdot v_S^{0.5} \cdot N^{0.5}$$

For explanation of symbols see previous page.

This equation exhibits a good correlation for volumes up to 60 l. Liquid density, viscosity, and surface tension should be constant.

In the case of pilot and production plant fermentors, other constants and exponents have been calculated; *K* and *a* will decrease with increasing V_0. Therefore, this equation is suitable for the determination of oxygen transfer efficiencies. For scaling-up procedures there are other more accurate relationships available which should be referred to.

Determination of $k_L \cdot a$ in aerated fermentors equipped with disc turbine impellers

$$k_L \cdot a = K' \left[t_M{}^{0.203} \left(\frac{P_G}{V_0} \right)^{1.79} \cdot M_f{}^{1.06} \cdot N^{0.0437} \right]$$

K' = constant $[s^4 \cdot m^{-3} \cdot kg^{-1}]$

t_M = mixing time [s]

P_G/V_0 = power input $[W \cdot m^{-3}]$ (cf. page 117)

M_f = momentum factor $[m^4 \cdot s^{-2}]$ (cf. page 121)

N = stirrer speed $[s^{-1}]$

The mixing time (t_m) must be experimentally determined.

Determination of fermentor geometry for given height–diameter ratio

for $H_L = x \cdot D_i$

$$H_L = 1.08 \cdot x^{\frac{2}{3}} \cdot V^{\frac{1}{3}}$$

$$D_i = 1.08 \cdot x^{-\frac{1}{3}} \cdot V^{\frac{1}{3}} = \frac{H_L}{x}$$

H_L = height of fermentor [m]

D_i = diameter of fermentor [m]

x = height–diameter ratio

V = fermentor volume [m^3]

The effective working volume of the fermentor (V_0) is less than V since head space and internal installations such as cooling coils and baffles require a certain volume themselves. The form of the fermentor construction such as concave end plates introduces a further slight error.

Reynolds number for impellers

$$Re = \frac{D_i^2 \cdot N \cdot \rho}{\eta}$$

Re = Reynolds number of conventionally agitated vessels (CSTR, Continuously Stirred Tank Reactor)

D_i = impeller diameter [m]

N = impeller speed [s^{-1}]

ρ = liquid density [$kg \cdot m^{-3}$]

η = dynamic viscosity of the medium [$N \cdot s \cdot m^{-2}$]

This equation describes Reynolds' law of similarity for impellers; the Reynolds number expresses the relationship between inertial and frictional (viscosity) forces arising due to mixing effects in Newtonian liquids.

In technical fermentations or bioprocesses the Reynolds number is closely correlated to the power number (N_P) over a wide range (cf. page 119).

In the immediate vicinity of the impeller, turbulent flow is present when Re is equal to or above the value of 10^3.

Synonyms: $Re = N_{Re}$

Froude number for impellers

$$Fr = \frac{N^2 \cdot D_i}{g}$$

Fr = Froude number for impellers

N = impeller speed [s^{-1}]

D_i = impeller diameter [m]

g = acceleration due to gravity

This equation describes Froude's law of similarity for impellers.

Froude's law expresses the relationship between inertial and gravitational forces. In fermentation technology hydrodynamic behaviour is similar in systems with the same Froude number. Fr is also used to describe the flooding characteristics of different impellers and the formation of vortexes in mixed systems.

Synonyms: $Fr = N_{Fr}$

Prandtl number for fluids

$$Pr = \frac{\eta \cdot c_P}{\lambda}$$

Pr = Prandtl number for fluids

η = dynamic viscosity $[N \cdot s \cdot m^{-2}]$

c_P = specific heat $[kJ \cdot kg^{-1} \cdot K^{-1}]$

λ = coefficient of thermal conductivity $[W \cdot m^{-1} \cdot K^{-1}]$

This equation describes Prandtl's law of similarity. The application of this law is related to the determination of internal friction and heat conductivity of flowing fluids. *Pr* depends on the temperature and nature of the fluid (liquid or gas).

At 20 °C *Pr* for water is 7.03; for air 0.71; for CO_2 0.84. For steam at 100 °C, *Pr* is 0.99.

The Prandtl number is the quotient of two complex numbers:

$$Pr = \frac{\text{Péclet number}}{\text{Reynolds number}}$$

Schmidt and Sherwood numbers

$$Sc = \frac{\nu}{D} \qquad Sh = \frac{k_L \cdot d}{D}$$

ν = kinematic viscosity $[m^2 \cdot s^{-1}]$

D = diffusion coefficient $[m^2 \cdot s^{-1}]$

k_L = mass-transfer coefficient $[m \cdot s^{-1}]$

d = thickness of flowing liquid layer [m]

The Schmidt number (*Sc*) – formerly called Prandtl number of diffusion – compares the molecular values of impulse and material transport. *Sc* for a mixture of steam and air has a value of approximately 0.6; for a mixture of ethanol and water, *Sc* is approximately 1000. *Sc* is the quotient of the Bodenstein and Reynolds numbers. In fermentation technology, the *Bodenstein* number is used for the characterization of the effects of back-mixing.

The Sherwood number (*Sh*) describes the relationship between densities of mass transfer due to convection and diffusion. *Sh* can be employed both for mass transfer (cf. above) as well as for absorption processes.

Nusselt number

$$Nu = \frac{\alpha \cdot d}{\lambda}$$

Nu = Nusselt number

α = heat-transmission coefficient $[W \cdot m^{-2} \cdot K^{-1}]$

d = thickness of the layer of material [m]

λ = coefficient of thermal conductivity $[W \cdot m^{-1} \cdot K^{-1}]$

The Nusselt number can be described as being a "dimensionless heat-transmission coefficient" and characterizes the ratio of surface power density and heat conductivity for a layer with the thickness d. In some cases, Nu is correlated with the heat transfer coefficient (k instead of α, cf. p. 126). For stirred tank reactors, the relationships are as follows:

1. Stirred tanks with double-walled heat exchangers:

$$Nu = 0.36 \cdot Re^{0.67} \cdot Pr^{0.33} \cdot H_L^{0.14}$$

(Re 300–600 000; Pr 2–2 400; H_L up to 20 m)

2. Stirred tanks with internal cooling coil-type of heat exchangers:

$$Nu = 0.87 \cdot Re^{0.62} \cdot Pr^{0.33} \cdot H_L^{0.14}$$

The ratio of the internal fermentor diameter to that of the impeller is approximately 2 in both cases.

Centrifugal force and acceleration factor

$$F_C = m \cdot r \cdot \omega^2$$

$$f_C = \frac{r \cdot \omega^2}{g} = \frac{4\pi^2 r \cdot N^2}{g} \approx \frac{r \cdot (\text{rpm})^2}{900}$$

F_C = centrifugal force [N]

m = mass of a particle in the gravitational field [kg]

f_C = centrifugal factor; determines the multiplication factor of acceleration due to gravity

ω = angular velocity [s^{-1}]

r = distance of the particle to the axis of rotation [m]

N = speed of rotor, [s^{-1}] or revolutions per minute [rpm]

In the case of both fixed-angle and swing-out bucket types of rotors, r is not constant but increases during movement of the particle. Due to the geometry of the rotor, average, minimum and maximum values for f_C can be determined.

Further reading

1. **Atkinson B, Mavituna F**
Biochemical Engineering and Biotechnology Handbook.
Macmillan, 1983

2. **Considine D M**
Process Instruments and Controls Handbook 3rd edn.
McGraw Hill, New York, 1985

3. **Himmelblau D M**
Basic Principles and Calculations in Chemical Engineering 4th edn.
Prentice–Hall, Inc., Englewood Cliffs, N.J., 1982

4. **Martin A N, Swarbrick J, Cammarata A**
Physikalische Pharmazie.
Wissenschaftliche Verlagsgesellschaft mbH, Stuttgart, 1975

5. **Aiba S, Humphrey A E, Millis N E**
Biochemical Engineering 2nd edn.
Academic Press, Inc., New York, 1973

6. **Roels J A**
Energetics and Kinetics in Biotechnology.
Elsevier Biomedical Press, Amsterdam, 1983

7. **Bailey J E, Ollis D F**
Biochemical Engineering Fundamentals 2nd edn.
McGraw-Hill, 1986

8. **Moo-Young M**
Comprehensive Biotechnology vol. 2.
Pergamon Press, 1985

9. **Lehninger A L**
Biochemistry.
Worth Publishers, Inc.

10. **Stryer L**
Biochemistry 2nd edn.
Freeman and Company, San Francisco, 1981

11. **Wang D I C, Cooney C L, Demain A L, Dunhill P, Humphrey A E, Lilly M D**
Fermentation and Enzyme Technology.
John Wiley & Sons, New York, 1979

12. **Stanbury P F, Whitaker A**
Principles of Fermentation Technology.
Pergamon Press, Oxford, 1984

APPENDIX

Table 1: Physical constants

Gravity:	$g = 9.81\ m \cdot s^{-2}$
Molar volume:	$V_M = 22.41\ l \cdot mol^{-1}$
Avogadro's constant:	$N = 6.02 \cdot 10^{23}\ mol^{-1}$
Faraday's constant:	$F = 9.65 \cdot 10^4\ A \cdot s \cdot mol^{-1}$
Planck's constant:	$h = 6.63 \cdot 10^{-34}\ J \cdot s$
Velocity of light:	$c = 3 \cdot 10^8\ m \cdot s^{-1}$
Elementary charge:	$e = 1.602 \cdot 10^{-19}\ A \cdot s$
Mass of an electron at rest:	$m_e = 9.109 \cdot 10^{-31}\ kg$
Gas constant,	
– general:	$R = 8.31\ J \cdot mol^{-1} \cdot K^{-1}$
– for air:	$R_A = 0.288\ J \cdot g^{-1} \cdot K^{-1}$
– for oxygen	$R_{O_2} = 0.260\ J \cdot g^{-1} \cdot K^{-1}$
– for CO_2:	$R_{CO_2} = 0.186\ J \cdot g^{-1} \cdot K^{-1}$
– for nitrogen:	$R_{N_2} = 0.297\ J \cdot g^{-1} \cdot K^{-1}$

Table 2: SI units

Basic Units

Variable		Measuring unit	
Term	**Symbol**	**Term**	**Symbol**
Length	l	meter	m
Time	t	second	s
Mass	m	kilogram	kg
Electric current	I	ampere	A
Thermodynamic temperature	T	kelvin	K
Amount of substance	n	mole	mol
Luminous intensity	l_v	candela	cd

Dimensions and Mechanics

Variable		Measuring unit		
Term	**Symbol**	**Term**	**Symbol**	**SI system**
Area	A	–	–	m^2
Volume	V	–	–	m^3
Angle (two-dimensional)	ϕ	radian	rad	$m \cdot m^{-1}$
Solid angle	Ω	steradian	sr	$m^2 \cdot m^{-2}$
Velocity	v	–	–	$m \cdot s^{-1}$
Angular velocity	ω	–	–	$rad \cdot s^{-1}$ (simple form: s^{-1})
Acceleration	a	–	–	$m \cdot s^{-2}$
Angular acceleration	α	–	–	$rad \cdot s^{-2}$ (simple form: s^{-2})
Speed (number of revolutions)	n	–	–	s^{-1}
Frequency	f	hertz	Hz	s^{-1}
Angular frequency	ω	–	–	$rad \cdot s^{-1}$ (simple form: s^{-1})
Force	F	newton	N	$kg \cdot m \cdot s^{-2}$
Density	ρ	–	–	$kg \cdot m^{-3}$
Momentum	p	–	–	$kg \cdot m \cdot s^{-1}$
Angular momentum	L	–	–	$kg \cdot m^2 \cdot s^{-1}$
Moment of inertia	J	–	–	$kg \cdot m^2$
Torque	M	–	–	$N \cdot m$
Pressure	P	pascal	Pa	$N \cdot m^{-2} = kg \cdot m^{-1} \cdot s^{-2}$
Dynamic viscosity	η	–	–	$Pa \cdot s = kg \cdot m^{-1} \cdot s^{-1}$
Kinematic viscosity	v	–	–	$m^2 \cdot s^{-1}$
Work, energy	W	joule	J	$N \cdot m = W \cdot s$
Power	P	watt	W	$J \cdot s^{-1} = V \cdot A = N \cdot m \cdot s^{-1}$

Thermodynamics

Variable		Measuring unit		
Term	Symbol	Term	Symbol	SI system
Quantity of heat enthalpy	Q, H	joule	J	–
Heat capacity	C_v	–	–	$J \cdot K^{-1}$
Entropy	S	–	–	$J \cdot K^{-1}$
Surface power density	ϕ	–	–	$W \cdot m^{-2}$
Heat transmission coefficient	α	–	–	$W \cdot m^{-2} \cdot K^{-1}$
Heat transfer coefficient	k	–	–	$W \cdot m^{-2} \cdot K^{-1}$
Thermal conduction coefficient	λ	–	–	$W \cdot m^{-1} \cdot K^{-1}$

Electricity and Magnetism

Variable		Measuring unit		
Term	Symbol	Term	Symbol	Unit term
Charge	Q	coulomb	C	$A \cdot s$
Voltage	V	volt	V	$J \cdot C^{-1} = Wb \cdot s^{-1}$
Field strength	E	–	–	$V \cdot m^{-1} = N \cdot C^{-1}$
Magnetic flux	Φ	weber	Wb	$V \cdot s = T \cdot m^2$
Magnetic flux density	B	tesla	T	$Wb \cdot m^{-2}$
Resistance	R	ohm	Ω	$V \cdot A^{-1} = S^{-1}$
Specific resistance	ρ	–	–	$\Omega \cdot m$
Conductance	G	siemens	S	$A \cdot V^{-1} = \Omega$
Inductance	L	henry	H	$Wb \cdot A^{-1} = \Omega \cdot s$
Capacitance	C	farad	F	$C \cdot V^{-1} = S \cdot s$

Table 3: Conversion tables

Temperature

Units	°C	K	°F	°R
t_c* °Celsius	t_c	$t_c+273.15$	$1.80\, t_c+32$	$180\, t_c+491.67$
T_K* Kelvin	$T_K-273.15$	T_K	$1.80(T_K-273.15)+32$	$1.80\, T_K$
t_F* °Fahrenheit	$0.5556)t_F-32)$	$0.5556(t_F-32)+273.15$	t_F	$t_F+459.7$
T_R* °Rankine	$0.5556(T_R-491.67)$	$0.5556\, T_R$	$T_R-459.67$	T_R

* For converting t_c, T_K, t_F, and T_R the corresponding values should be used, for example 7° C=(7+273.15) K=280.15 K

Pressure

Unit	Pa (N/m²)	bar	kp/cm² = at	atm
1 Pa = 1 N/m²	1	10^{-5}	$1.01972 \cdot 10^{-5}$	$0.98692 \cdot 10^{-5}$
1 bar	10^5	1	1.01972	0.98692
1 kp/cm² = 1 at	$0.98067 \cdot 10^5$	0.98067	1	0.96784
1 atm	$1.01325 \cdot 10^5$	1.01325	1.0332	1
1 torr	$1.3332 \cdot 10^2$	$1.3332 \cdot 10^{-3}$	$1.3595 \cdot 10^{-3}$	$1.31579 \cdot 10^{-3}$
1 mWC	$0.98066 \cdot 10^4$	$0.98067 \cdot 10^{-1}$	0.1	$0.96784 \cdot 10^{-1}$
1 mmWC = 1 kp/m²	9.8066	$0.98066 \cdot 10^{-4}$	10^{-4}	$0.96784 \cdot 10^{-4}$

Unit	torr	mWC*	mmWC* = kp/m²
1 Pa = 1N/m²	$7.5006 \cdot 10^{-3}$	$1.01972 \cdot 10^{-4}$	$1.01972 \cdot 10^{-1}$
1 bar	$7.5006 \cdot 10^2$	$1.01972 \cdot 10$	$1.01972 \cdot 10^4$
1 kp/cm² = 1 at	$7.3556 \cdot 10^2$	10	10^4
1 atm	$7.60 \cdot 10^2$	$1.0332 \cdot 10$	$1.0332 \cdot 10^4$
1 torr	1	$1.3595 \cdot 10^{-2}$	$1.3595 \cdot 10$
1 mWC	$7.3556 \cdot 10$	1	10^3
1 mmWC = 1 kp/m²	$7.3556 \cdot 10^{-2}$	10^{-3}	1

* WC = water column

Power

Unit	W (J/s, Nm/s)	kpm/s	kcal/h	PS*
1 W (J/s, Nm/s)	1	0.1019716	0.859834	$1.3596 \cdot 10^{-3}$
1 kpm/s	9.80665	1	8.43210	$1.3333 \cdot 10^{-2}$
1 kcal/h	1.163	0.118593	1	$1.5812 \cdot 10^{-3}$
1 PS	$7.35499 \cdot 10^{2}$	75	$6.3241 \cdot 10^{2}$	1

* PS = metric horsepower

Energy

Unit	J (Nm, Ws)	kpm	kcal
1 J (Nm, Ws)	1	0.10197	$2.3885 \cdot 10^{-4}$
1 kpm	9.80665	1	$2.3423 \cdot 10^{-3}$
1 kcal	$4.18684 \cdot 10^{3}$	$4.26935 \cdot 10^{2}$	1
1 kWh	$3.6 \cdot 10^{6}$	$3.67098 \cdot 10^{5}$	$8.59845 \cdot 10^{2}$
1 PSh	$2.6478 \cdot 10^{6}$	$2.7 \cdot 10^{5}$	$6.3242 \cdot 10^{2}$
1 erg	10^{-7}	$0.10197 \cdot 10^{-7}$	$2.388 \cdot 10^{-11}$

Unit	kWh	PSh	erg
1 J (Nm, Ws)	$2.7778 \cdot 10^{-7}$	$3.7767 \cdot 10^{-7}$	10^{7}
1 kpm	$2.7241 \cdot 10^{-6}$	$3.70370 \cdot 10^{-6}$	$9.8066 \cdot 10^{7}$
1 kcal	$1.163 \cdot 10^{-3}$	$1.5812 \cdot 10^{-3}$	$4.1868 \cdot 10^{10}$
1 kWh	1	1.35962	$3.6 \cdot 10^{13}$
1 PSh	0.7355	1	$2.6478 \cdot 10^{13}$
1 erg	$2.777 \cdot 10^{-14}$	$3.776 \cdot 10^{-14}$	1

Kinematic Viscosity

Unit	m^2/s	m^2/h	$St = cm^2/s$
1 m^2/s	1	$3.600 \cdot 10^{3}$	10^{4}
1 m^2/h	$2.7778 \cdot 10^{-4}$	1	2.7778
1 St = 1 cm^2/s	10^{-4}	0.36	1

Specific Heat

Unit	J/kg K	kcal/kg K = cal/g K	kWh/kg K
1 J/kg K	1	$2.388 \cdot 10^{-4}$	$2.777 \cdot 10^{-7}$
1 kcal/kg K = 1 cal/g K	$4.1868 \cdot 10^{3}$	1	$1.163 \cdot 10^{-3}$
1 kWh/kg K	$3.6 \cdot 10^{6}$	$8.59845 \cdot 10^{2}$	1

Thermal Conductivity Coefficients

Unit	W/m K	kW/m K	W/cm K	kcal/m h K
1 W/m K	1	10^{-3}	10^{-2}	0.8598
1 kW/m K	10^{3}	1	10	$8.598 \cdot 10^{2}$
1 W/cm K	10^{2}	0.1	1	8.598
1 kcal/m h K	1.163	$1.163 \cdot 10^{-3}$	$1.163 \cdot 10^{-2}$	1

Heat Transfer Coefficients, Heat Transmission Coefficients

Unit	W/m² K	kW/m² K	W/cm² K	kcal/m² h K
1 W/m² K	1	10^{-3}	10^{-4}	0.859845
1 kW/m² K	10^{3}	1	0.1	$8.59845 \cdot 10^{2}$
1 W/cm² K	10^{-4}	10	1	$8.59845 \cdot 10^{3}$
1 kcal/m² h K	1.163	$1.163 \cdot 10^{-3}$	$1.163 \cdot 10^{-4}$	1

Table 4: Prefixes for powers of ten

Power of ten	Prefix	Prefix symbol
10^{18}	exa	E
10^{15}	peta	P
10^{12}	tera	T
10^{9}	giga	G
10^{6}	meta	M
10^{3}	kilo	k
10^{2}	hecto	h
10	deca	da
10^{-1}	deci	d
10^{-2}	centi	c
10^{-3}	milli	m
10^{-6}	micro	μ
10^{-9}	nano	n
10^{-12}	pico	p
10^{-15}	femto	f
10^{-18}	atto	a

Table 5: English and Greek alphabets

English alphabet		Greek alphabet			
A	a	A	α	a	alpha
B	b	B	β	b	beta
C	c	Γ	γ	g	gamma
D	d	Δ	δ	d	delta
E	e	E	ε	e	epsilon
F	f	Z	ζ	z	zeta
G	g	H	η	e	eta
H	h	Θ	θ	th	theta
I	i	I	ι	i	iota
J	j	K	κ	k	kappa
K	k	Λ	λ	l	lambda
L	l	M	μ	m	mu
M	m	N	ν	n	nu
N	n	Ξ	ξ	x	xi
O	o	O	o	o	omicron
P	p	Π	π	p	pi
Q	q	P	ρ	r	rho
R	r	Σ	σ	s	sigma
S	s	T	τ	t	tau
T	t	Y	υ	u	upsilon
U	u	Φ	ϕ	ph	phi
V	v	X	χ	ch	chi
W	w	Ψ	ψ	ps	psi
X	x	Ω	ω	o	omega
Y	y				
Z	z				

Table 6: Table of vapour pressures

Temperature t (°C)	Pressure p (bar)	Specific steam volume v'' (m^3/kg)	Steam density ρ'' (kg/m^3)	Specific enthalpy Water h' (kJ/kg)	Specific enthalpy Steam h'' (kJ/kg)
0	0.006108	206.3	0.004846	0	2500
10	0.012271	106.4	0.009396	42.04	2519
20	0.023692	57.84	0.01729	83.86	2537
30	0.042414	32.93	0.03036	125.6	2556
40	0.073760	19.55	0.05114	167.4	2574
50	0.123348	12.05	0.08298	209.1	2592
60	0.19917	7.682	0.1302	251.0	2609
70	0.31157	5.049	0.1981	292.8	2626
80	0.47356	3.410	0.2933	334.7	2643
90	0.70108	2.361	0.4235	376.7	2659
100	1.01325	1.673	0.5977	418.8	2675
110	1.4326	1.210	0.8265	461.1	2690
120	1.9854	0.8914	1.122	503.7	2705
130	2.7011	0.6680	1.496	546.0	2718
140	3.6137	0.5084	1.967	588.7	2732
150	4.7601	0.3924	2.548	631.8	2744
160	6.1802	0.3068	3.260	675.3	2756
170	7.9198	0.2426	4.122	718.9	2767
180	10.027	0.1939	5.157	762.8	2777
190	12.553	0.1564	6.392	807.2	2785
200	15.550	0.1273	7.857	852.0	2793
250	39.776	0.05006	19.98	1085	2801
300	85.916	0.02163	46.24	1344	2747
350	165.37	0.008803	113.6	1670	2562
374.2	221.14	0.00314	319	2102	2102

Table 7: Relative atomic masses (selected elements)

Aluminium	Al	26.98	Nickel	Ni	58.7
Antimony	Sb	121.7	Niobium	Nb	92.91
Argon	Ar	39.94	Nitrogen	N	14.01
Arsenic	As	74.92	Osmium	Os	190.2
Barium	Ba	137.3	Oxygen	O	16.00
Beryllium	Be	9.01	Palladium	Pd	106.4
Bismuth	Bi	208.98	Phosphorus	P	30.97
Boron	B	10.81	Platinum	Pt	195.0
Bromine	Br	79.90	Potassium	K	39.09
Cadmium	Cd	112.40	Protactinium	Pa	231.04
Carbon	C	12.01	Radium	Ra	226.02
Caesium	Cs	132.90	Rhodium	Rh	102.91
Chlorine	Cl	35.45	Rubidium	Rb	85.47
Chromium	Cr	52.00	Scandium	Sc	44.96
Cobalt	Co	58.93	Selenium	Se	78.9
Copper	Cu	63.54	Silicon	Si	28.08
Fluorine	F	19.00	Silver	Ag	107.87
Gallium	Ga	67.72	Sodium	Na	22.99
Germanium	Ge	72.5	Strontium	Sr	87.62
Gold	Au	197.00	Sulphur	S	32.06
Helium	He	4.00	Tantalum	Ta	180.95
Hydrogen	H	1.01	Technetium	Tc	98.91
Indium	In	114.82	Tellurium	Te	127.6
Iodine	I	126.90	Thallium	Tl	204.3
Iridium	Ir	192.2	Thorium	Th	232.04
Iron	Fe	55.84	Tin	Sn	118.6
Krypton	Kr	83.80	Titanium	Ti	47.9
Lanthanum	La	138.90	Uranium	U	238.03
Lead	Pb	207.2	Vanadium	V	50.94
Lithium	Li	6.94	Wolfram (tungsten)	W	183.8
Magnesium	Mg	24.30			
Manganese	Mn	54.94	Xenon	Xe	131.30
Mercury	Hg	200.59	Ytterbium	Yb	173.0
Molybdenum	Mo	95.9	Zinc	Zn	65.38
Neon	Ne	20.17	Zirconium	Zr	91.22

Reference element $^{12}C = 12.000$

The Periodic Table

Period-number (electron shell)		Group number I	II						
1 (K)		1 H 1.008							
2 (L)		3 Li 6.9	4 Be 9.0						
3 (M)		11 Na 23.0	12 Mg 24.3						
4 (N)		19 K 39.1	20 Ca 40.1	21 Sc 44.9	22 Ti 47.9	23 V 50.9	24 Cr 51.9	25 Mn 54.9	26 Fe 55.8
5 (O)		37 Rb 85.5	38 Sr 87.6	39 Y 88.9	40 Zr 91.2	41 Nb 92.9	42 Mo 95.9	43 Tc (97.0)	44 Ru 101.7
6 (P)		55 Cs 132.9	56 Ba 137.4	57 La 139 Lanthanides	72 Hf 178.6	73 Ta 180.9	74 W 183.8	75 Re 186.3	76 Os 190.2
7 (Q)		87 • Fr 223.0	88 • Ra 226.0	89 • Ac 227 Actinides	104 • Ku 267				

Rare earth metals (Lanthanides)	58 Ce 140.1	59 Pr 140.9	60 Nd 144.2	61 • Pm 145	62 Sm 150

Transuranic elements (Actinides)	90 • Th 232.1	91 • Pa 231	92 • U 238	93 • Np 237	94 • Pu 244

Atomic masses in relation to 1/12 of the mass of ^{12}C (reference element);
Atomic masses in parenthesis: mass of the isotope with the longest life

				III	IV	V	VI	VII	O
									2 He 4.0
				5 B 10.8	6 C 12.0	7 N 14.0	8 O 16.0	9 F 19.0	10 Ne 20.2
				13 Al 26.9	14 Si 28.1	15 P 31.0	16 S 32.0	17 Cl 35.5	18 Ar 39.9
27 Co 58.9	28 Ni 58.7	29 Cu 63.5	30 Zn 65.4	31 Ga 69.7	32 Ge 72.6	33 As 74.9	34 Se 78.9	35 Br 79.9	36 Kr 83.8
45 Rh 102.9	46 Pd 106.4	47 Ag 107.9	48 Cd 112.4	49 In 114.8	50 Sn 118.7	51 Sb 121.7	52 Te 127.6	53 J 126.9	54 Xe 131.3
77 Ir 192.2	78 Pt 195.1	79 Au 196.9	80 Hg 200.6	81 Tl 204.4	82 Pb 207.2	83 Bi 208.9	84 Po (209.0)	85 At (210.0)	86 Rn (222.0)
								Metals	Non-metals

63 Eu 151.9	64 Gd 157.2	65 Tb 158.9	66 Dy 162.5	67 Ho 164.9	68 Er 167.2	69 Tm 168.9	70 Yb 173	71 Lu 174.9	

95 • Am 243	96 • Cm 247	97 • Bk 247	98 • Cf 251	99 • Es 254	100 • Fm 253	101 • Md 256	102 • No 256	103 • Lw 257	

• All isotopes of this element are radioactive

Subject index